Variety of Development

Qiushi Feng

Variety of Development

Chinese Automakers in Market Reform and Globalization

palgrave
macmillan

Qiushi Feng
Department of Sociology
National University of Singapore
Singapore, Singapore

ISBN 978-981-10-5911-7 ISBN 978-981-10-5912-4 (eBook)
https://doi.org/10.1007/978-981-10-5912-4

Library of Congress Control Number: 2017955053

Printed on acid-free paper

This Palgrave imprint is published by Springer Nature
The registered company is Springer Nature Singapore Pte Ltd.
The registered company address is: 152 Beach Road, #21-01/04 Gateway East, Singapore 189721, Singapore

Preface

The idea for this book originated in the early years of my doctoral studies at the Department of Sociology, Duke University. For these initial stages of this research, I deeply appreciate the guidance of my teachers during the Duke years, especially members of my dissertation committee, Professors Bai Gao, Nan Lin, Gary Gereffi, Edward A. Tiryakian, and David W. Soskice. I was deeply fortunate to know these senior scholars in my early career, who not only lead their respective fields, but were always generous and warm-hearted mentors.

I would like to give special thanks to my advisor, Professor Bai Gao, who has been in many ways an inspiration. Without his insightful instruction, sharp comment, and constant encouragement, this project may never have been accomplished. What I have learned from him is not only specific views or approaches in sociology, but ways of observing the world and thinking in the social sciences. I would like to thank classmates on my PhD program Wei Zhao, Li-Hsuan Cheng, and Ryan Denniston, who also provided valuable comments and suggestions in the early stages of this book project. I also wish to thank Duke Sociology Department for its wonderful PhD program and generous financial support, the Asia/Pacific Studies Institute of Duke for the summer fieldwork fellowship, and the Duke Graduate School for the Julian Prices Fellowship, all of which greatly facilitated my research on the Chinese automobile industry.

I remain indebted to many people in relation to my fieldwork in China. The advisor for my master's thesis, Professor Meng Xianzhong in Shanghai Jiao Tong University, gave me significant support and has been a constant guide both in academia and throughout my life. Professor Bing Zheng in Jilin University, my college teacher in sociology, never hesitated to help me during fieldwork whenever needed. Many friends supported me in the data collection for this project. I especially thank Wang Huiqun, Zhen Zhihong, Xin Benlu, Dong Yunsheng, Liu Jianwei, Shen Yuliang, Xu Jialiang, and Fang Qing. With their support, I gained precious opportunities to observe and analyze how the Chinese automobile industry was functioning in the recent boom years.

I started work on the manuscript for this book after joining the Department of Sociology, National University of Singapore. In recent years, numerous department colleagues have provided important comments on this project, including Noorman Abdullah, Chau Beng Huat, Vincent Chua, Kurtulus Gemici, Jiwook Jung, Kelvin Low, Manjusha Nair, Anne Raffin, Vineeta Sinha, Xiaohong Xu, and Jean W. Yeung. Special thanks also to Joonmo Son, an old Duke friend and now a department colleague in Singapore, and Salvatore Babones at the University of Sydney, both of whom have given a number of critical suggestions for the work.

I would like to thank Jacob Dreyer, my editor at Palgrave Macmillan, who has provided great support for the project, and Jazmine Robles for professional assistance in book production. I greatly appreciate the help of Robert Cole at the Department of Geography, National University of Singapore, for his exceptional proofreading and assistance in preparing the final draft. I also wish to acknowledge the book grant provided by the Faculty of Arts and Social Sciences, National University of Singapore, which facilitated this publication.

Finally, my family deserves much credit in this project. When I started my doctoral studies, my wife, Zhao Mian, abandoned her career in Japan and moved to Duke to support me without hesitation. While providing me the best partnership through the Duke years, she also worked hard to gain a second master's degree at the University of North Carolina at Chapel Hill. When we later moved to Singapore, the loud arrival in the world of our dear son, Feng Yanbo, proved a highly effective motivation for me to complete this book. My parents-in-law, Zhao Guang and Xing Tianyun, showed their sincere trust in me by sending their daughter to America when I started my academic journey overseas. My parents, Feng Zhanji and Gou Zhiying, are quite simply the best parents one could hope for in this world. Remembering their visible anxiety one summer's day in

2002 as I was about to take a flight to the United States, I hope this book may provide a long-due answer to them. Their support in my life is an integral part of this book, though only readable for me. I would also like to thank my grandfather, Feng Shouji, who will soon celebrate his ninetieth birthday. This book, I hope, will be a gift for him. Now with the book in hand, I hope to be less burdensome towards my family in the near future, by not returning home so late at night.

Qiushi Feng

Contents

Abbreviations

BAIC	Beijing Automobile Industry Holding Corporation
CKD	Complete knock-down
CNAIC	China National Automobile Industry Corporation
CPC	Communist Party of China
DC	Daimler Chrysler
DMC	Dongfeng Motor Corporation
FAW	First Auto Works
FDI	Foreign Direct Investment
GAIG	Guangzhou Automobile Industry Group Corporation
GDP	Gross Domestic Product
GM	General Motors
IMF	International Monetary Fund
JAC	Anhui Jianghuai Automobile Co.
MPV	Multi-purpose vehicle
NAFTA	North American Free Trade Agreement
OBM	Original brand manufacture
OEM	Original equipment manufacture
OICA	Organisation Internationale des Constructeurs d'Automobiles
PSA	PSA Peugeot Citroën
R&D	Research and Development
SAC	Shanghai Automotive Corporation
SAIC	Shanghai Automobile Industry Corporation
SASAC	State-owned Asset Supervision and Administration Commission

SAW	Second Auto Works
STAICC	Shanghai Tractor and Automobile Industry Coordination Co.
SUV	Sport utility vehicle
VW	Volkswagen
WCIC	Wuhu Construction and Investment Corporation
WTO	World Trade Organization

List of Figures

List of Tables

CHAPTER 1

Introduction

The meteoric rise of China in the global economy over the past 30 years, and the far-reaching effects of this on the social, political, and economic landscape of the country, have attracted wide attention throughout the social sciences. Since the start of the market reform in the early 1980s, China's Gross Domestic Product (GDP) has maintained average annual growth of around 10%, catapulting the economy to become the second largest in the world. This dramatic shift was greatly facilitated by China's entry to the World Trade Organization (WTO) in 2001. Within 10 years of this historic development, China had become the world's greatest merchandise exporter, while simultaneously attracting the largest total Foreign Direct Investment (FDI) of all developing countries.

One of the best illustrations of the complex and often contradictory story of Chinese economic development can be found in the automobile sector. Consuming about half the world's oil, the highly capital- and technology-intensive automobile industry is among the most significant economic sectors at the global level. At national level, this sector often plays a critical role in domestic economies, as it typically comprises long and heavily interdependent supply and production networks, managed through intricate coordination systems. For developing countries, the prevalence of car use for personal transport is a good indicator of development, industrialization, and consumer purchasing power. All of these considerations ring true in the context of China's economic transition.

Q. Feng, *Variety of Development*,
https://doi.org/10.1007/978-981-10-5912-4_1

The Rise of the Chinese Automobile Industry

China is getting on wheels. In recent memory, for most Chinese bicycles were the mainstay of personal transport, but by the end of 2003 they were banned on Shanghai's main avenues to make way for the rapidly expanding ownership and use of cars. Over the past three decades, with unparalleled speed, China has emerged as the world's largest producer and consumer of automobiles. During the planned-economy era, from the 1950s to 1980s, China's total annual automobile output never reached more than half a million vehicles per year. A complete reversal of this trend took place in the 1990s, gathering pace towards the turn of the new millennium. As can be seen in Table 1.1, China's automobile output increased almost 1400% in 16 years, from about 1.6 million vehicles in 1998 to about 23.7 million in 2014. As output accelerated, so China gradually moved up the rankings of global automakers, from tenth in the world in 1998, reaching the top five after China's achievement of WTO membership in 2001, and finally becoming the number one producer by 2009, when China's output totaled about 14 million, almost equivalent to that of Japan and the United States combined. By 2014, China's leading global position was firmly fixed, with a total output of 24 million automobiles, more than double that of the United States, and 2 million units ahead of the sum of the United States and Japan.

Passenger cars, mostly sedan models, hold a unique position in the exponential growth of Chinese automobile production. Prior to the market reform, cars occupied less than 1% of total national output, which was overwhelmingly geared towards functional vehicles such as trucks. The balance began to tip in the late 1990s, as the domestic consumer market started to expand and passenger car output increased to about 30% of total automobile production. Over the next ten years, cars continued to gain prominence. By 2006, about 70% of Chinese automobile output was accounted for by passenger cars, reaching 84% in 2014. This shows the clear role of passenger cars in the recent development trajectory of China's automobile sector.

The rise of the Chinese automobile sector, particularly the production of passenger cars, is firmly linked to the integration of China within the globalizing economy and the concurrent transformation of domestic institutions resulting from the market reform. The past three decades have seen China's automobile sector become ever more closely integrated with the world automobile industry. China's foreign trade in automobiles and automotive parts has been highly active since the late 1990s (Fig. 1.1). In 1998,

Table 1.1 World top ten automobile producing countries, 1998–2014

Rank	*1998*		*2001*		*2002*		*2006*		*2009*		*2014*	
	Country	*Output*	*Country*	*Output*	*Country*	*Output*	*Country*	*Output*	*Country*	*Output*	*Country*	*Output*
1	USA	12,006,079	USA	11,424,689	USA	12,279,582	Japan	11,484,233	China	13,790,994	China	23,731,600
2	Japan	10,049,792	Japan	9,777,191	Japan	10,257,315	USA	11,263,986	Japan	7,934,057	USA	11,660,702
3	Germany	5,726,788	Germany	5,691,677	Germany	5,469,309	China	7,188,708	USA	5,709,431	Japan	9,774,665
4	France	2,954,160	France	3,628,418	France	3,701,870	Germany	5,819,614	Germany	5,209,857	Germany	5,907,548
5	Spain	2,826,063	Korea	2,946,329	China	3,286,804	Korea	3,840,102	Korea	3,512,926	Korea	4,524,932
6	Canada	2,172,662	Spain	2,849,888	Korea	3,147,584	France	3,169,219	Brazil	3,182,923	India	3,844,857
7	UK	1,975,656	Canada	2,532,742	Spain	2,855,239	Spain	2,777,435	India	2,641,550	Mexico	3,368,010
8	Korea	1,954,494	China	2,334,440	Canada	2,629,437	Brazil	2,611,034	Spain	2,170,078	Brazil	3,146,386
9	Italy	1,692,737	Mexico	1,841,008	UK	1,823,018	Canada	2,572,292	France	2,047,693	Spain	2,402,978
10	China	1,627,829	Brazil	1,817,237	Mexico	1,804,670	Mexico	2,045,518	Mexico	1,561,052	Canada	2,394,154

Source: Statistics from Organisation Internationale des Constructeurs d'Automobiles (OICA) http://www.oica.net/category/production-statistics/

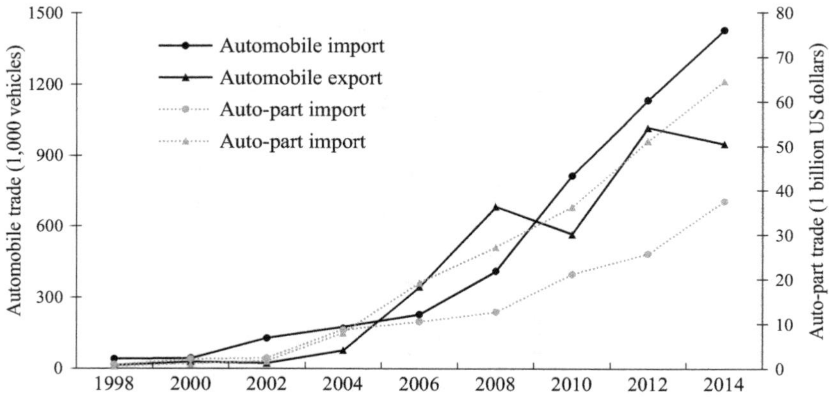

Fig. 1.1 Foreign trade in automobiles and parts in China, 1998–2014
Note: *Left axis* shows foreign trade in automobiles per 1000 units; *right axis* shows foreign trade in auto-parts per US$ 1 billion
Source China Automotive Industry Yearbooks

auto-parts imports and exports amounted to less than US$ 1 billion in China, but had soared to about US$ 100 billion by 2014. Regarding trade in complete vehicles, in 1998 China imported about 40 thousand automobiles and exported about 14 thousand; by 2012, imports and exports had both increased to around 1 million units.

Behind burgeoning automotive foreign trade were large inflows of FDI to the Chinese automobile sector from international automakers (Table 1.2). Since the American Motors Corporation, later taken over by Daimler Chrysler (DC), set up China's first joint venture to produce jeeps in 1983, and Volkswagen (VW) set up China's first joint venture for sedan cars in 1985, a succession of leading international automakers have swarmed into China. By 2004, major foreign manufacturers had mostly completed establishing their production arrangements in China. VW and General Motors (GM), being among the longest established investors in the Chinese auto sector, currently occupy the largest share of the market, though relative latecomers such as Ford and Toyota are fast catching up. Accompanying these foreign manufacturers were international suppliers, which followed on the heels of their clients to make inroads into the emerging Chinese market. Incoming

Table 1.2 Main assembly projects of transnational automobile corporations in China, 1983–2004

Automakers	*1983*	*1985*	*1991*	*1992*	*1993*	*1997*	*1998*	*2001*	*2002*	*2003*	*2004*
DC	●										●
VW		●	●								
PSA		●		●							
GM						●					
Honda							●			●	
Ford								●			
Toyota									●		●
Hyundai									●		
Kia									●		
Nissan										●	
BMW										●	

Source: Based on China Auto News (2005)

international corporations greatly facilitated the integration of China's automobile sector within global production networks.

From a domestic perspective, the development of China's automobile industry was driven by the sweeping institutional transformations put in place by the market reform. China's communist leaders had highlighted the importance of automobile production at the outset of the planned economy, considering the expansion of mechanized transport and related manufacturing capabilities indispensable to national industrialization. In consequence, the automobile industry was built as a national template for the essential institutional arrangements of the planned economy. Automobile enterprises were established as factories owned and directed by central or local governments, in accordance with hierarchical economic commands. The leading authority within the factories was the party committee, under which the directors were in charge of routine administration. With this governance structure, enterprises had minimal autonomy, and could be regarded as unthinking tools for fulfilling central government objectives.

However, this sector was heavily amended in the reform process. With respect to macro governance, the former centralized, top-down structure was largely abandoned. The central government initially attempted to organize domestic automakers under a large semi-corporatized association, but soon opted instead to release these enterprises as independent players in the emerging market environment. Industrial policies, instead of bureaucratic commands, became the principal measures through which the government extended its influence. Meanwhile, the central authority retained its grip on a few key automotive enterprises through ownership linkages. To facilitate their management, so-called "modern enterprise institutions" were introduced in place of the former routine administration by direct government intervention. In the process, these enterprises initially gained a degree of administrative autonomy from the state, and were later corporatized through a series of further institutional transformations, including clarification of their ownership structure and the establishment of boards of directors. Nowadays, though the state is an ever-present factor in different ways and in other national contexts, China's automobile sector exhibits many similarities to its Western counterparts in terms of the way it is governed (see State Council of China 1994a, b, 2004).

Given the impressive automobile production statistics, is China practicing a unique model of industry development which might be replicable elsewhere? Clearly, the recent torrent of production in China is much dependent on huge domestic demand, and thus on the sheer size of the Chinese population coupled with the growth of the national economy and consumer spending. However, it would be erroneous to suggest that China had no distinctive strategies in developing its automobile sector compared with other countries. As the "industry of industries", the automobile sector extends to many supply operations such as steel and electronics, and is therefore often used as a measure to stimulate broader economic development. But the huge investments and advanced technologies required for automobile projects are often in short supply in developing countries. This is particularly true for the production of passenger cars. Two major strategies are commonly deployed to overcome capital and technology constraints.

One such strategy is to attract foreign corporations to the domestic market, in other words, a model of dependence. Mexico's car industry provides a good example of this model (Bennett and Sharpe 1985). Before the 1960s, Mexico mainly imported automobiles from the United States, and the only domestic automobile manufacturing took the form of complete knock-down (CKD) production, in which all auto-parts were imported for local assembly. In the 1960s, the Mexican government attempted a policy of import substitution, setting high tariffs to block imports and requesting foreign automakers to set up joint ventures with domestic companies. However, under political pressure from the United States, American automobile giants such as Ford and GM ultimately dominated the local industry, wiping out national assemblers. Starting from the 1970s, export promotion became the new policy concern, which aligned well with the interests of international automakers, who would then employ Mexico as a production base for export to the American market. The establishment of the North American Free Trade Agreement (NAFTA) in 1994 further reinforced this arrangement. In consequence, Mexico was integrated into the global automobile production system under the control of large foreign corporations.

The second development strategy is to forge self-reliant national automakers through strong state interventions, or a model of independence, as is well represented by the South Korean example. In the 1970s, the Korean automobile industry began to take shape with Hyundai, Daewoo and Kia as the big three national manufacturers. Korean officials had a vision of building a competent and competitive national automobile industry, viewing foreign automakers as threats and strictly regulating FDI in the sector, while also harnessing the national banking system to finance homegrown automakers (Amsden 1989). In consequence, Hyundai, Daewoo and Kia all matured as national automakers independently of foreign corporations (Green 1992). With full government support, these firms became large conglomerates, the so-called chaebols. By the 1980s, the state was encouraging national automakers to look beyond the limited domestic market. Huge investments were channeled from the state to help build national automakers that could compete in the international market. A "national champion" strategy, through which official financial support was linked to the export performance of the enterprises, was also used by the government to incentivize exports. Since the 1990s, Korean national brands have established themselves as a major force in the global automobile market.

Though the two development strategies described above are readily identifiable in other contexts, it is hard to recognize either as a dominant model in China. To illustrate the characteristics of Chinese automobile production, Table 1.3 lists all domestic carmakers with sales of over 50,000 units in 2003. After two years of WTO membership, this specific point along the recent growth trajectory of China's automobile sector is an effective one to capture its development pattern. As can been seen, unlike in Mexico, there were no international automakers dominating the Chinese market. Although Volkswagen held a significant position, with

Table 1.3 Car makers in China with over 50,000 sales in 2003

Rank	*Name*	*Sales (unit)*	*Market share (%)*	*Joint venture*	*Major products*
1	Shanghai-VW	405,111	19.77	Yes	Foreign models
2	FAW-VW	297,995	14.55	Yes	Foreign models
3	Shanghai GM	201,188	9.82	Yes	Foreign models
4	Tianjin FAW Xiali	117,335	5.73	No	Foreign models
5	Guangzhou Honda	117,180	5.72	Yes	Foreign models
6	Dongfeng PSA	103,126	5.03	Yes	Foreign models
7	Changan Suzuki	100,018	4.88	Yes	Foreign models
8	Chery Auto	90,387	4.41	No	National models
9	Dongfeng Nissan	65,126	3.18	Yes	Foreign models
10	FAW Hainan Mazda	53,128	2.59	No	Foreign models
11	Beijing Hyundai	52,128	2.54	Yes	Foreign models
12	Geely Auto	51,741	2.53	No	National models
13	FAW Car	51,266	2.50	No	Foreign models
14	Dongfeng Yueda Kia	51,741	2.49	Yes	Foreign models
Total		1,710,937	83.52		

Source: China Automotive Industry Yearbook 2004

Note: *FAW* First Auto Works, *VW* Volkswagen

two joint ventures taking around 34% of the market share, most major carmakers in the list were joint ventures. Meanwhile, China was also distinct from the Korean independent track, in that domestic automakers mostly made foreign cars introduced by their international joint-venture partners. The prevalence of joint ventures is a major feature of the Chinese automobile industry, and may indicate a unique developmental model, in which foreign and domestic automakers cooperated in an emerging market. However, this model could not be easily replicated elsewhere. To gain a foothold in an overseas market, international automakers generally preferred direct exports, or to set up wholly owned corporations. The establishment of joint ventures has often been a compromise for the foreign firm, and indicates strong negotiating power on the part of the host country. Moreover, the joint-venture model does not capture the full picture of China's automobile industry. Two of the corporations in Table 1.3, Chery Auto and Geely Auto, were not affiliated with foreign automakers in joint ventures , and were making national-brand models. Such variation in production arrangements among Chinese automakers warrants further systematic investigation.

The Puzzle

The central inquiry of this book concerns the divergences in production organization within the Chinese automobile sector. Due to the capital- and technology-intensive nature of automobile manufacturing, the source of investments and acquisition of technology are two critical aspects of production arrangements for a developing country such as China. These two dimensions point to how the variations in production models among Chinese automakers can be examined and understood: put simply, who is making cars in China, and how. In the context of China, the former can be understood via the ownership structure of a given automaker. In the ongoing market transition, three major ownership structures have emerged among Chinese carmakers: (i) those owned by the central government; (ii) those owned by local governments; and (iii) those owned by private companies. In terms of acquiring technology, Chinese automakers have tended to pursue one of two main strategies: (i) establishing joint ventures to introduce foreign models; or (ii) producing national-brand cars through self-reliant research and development (R&D).

The choice between these pathways has driven the main bifurcations among Chinese automakers' production arrangements. Four extant production models are thus identifiable among Chinese carmakers: enterprises owned by the central government and producing foreign-brand cars through joint ventures (Model 1); enterprises owned by local governments and producing foreign-brand cars through joint ventures (Model 2); enterprises owned by local governments and making national-brand cars independently (Model 3); and privately owned enterprises making national-brand cars independently (Model 4). This classification is illustrated in Fig. 1.2, which groups Chinese carmakers into the four production models across the two dimensions of ownership structure and technological strategy.

The enterprises categorized in Fig. 1.2 represent the major automakers in China in the early 2000s, which may contain more than one corporation as listed in Table 1.3. These include the Chinese "Big Three" (three large enterprises that were prioritized by the central government to lead the automobile industry in the early period of reform): First Auto Works (FAW), Shanghai Automotive Industry (Group) Corporation (SAIC) and Dongfeng Motor Corporation (DMC), which are clustered at the upper left panel of Fig. 1.2. These three entities held most of the top-ranked car production corporations in Table 1.3. Shanghai VW and Shanghai GM,

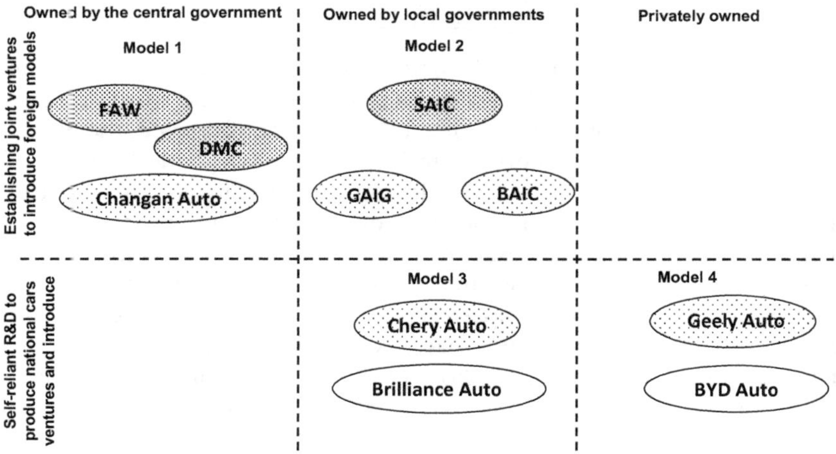

Fig. 1.2 Four production models of carmakers in China
Note: *Dot density* indicates three levels of vehicle sales by 2003: less than 200,000 units, 200,000 to 500,000 units and more than 500,000 units, respectively (*higher dot density* indicates higher sales). *FAW* First Auto Works, *DMC* Dongfeng Motor Corporation, *SAIC* Shanghai Automobile Industry Corporation, *GAIG* Guangzhou Automobile Industry Group Corporation, *BAIC* Beijing Automobile Industry Holding Corporation

respectively the largest and third-largest carmakers in Table 1.3, were under the umbrella of SAIC, nowadays the largest car manufacturer in China. Likewise, FAW-VW, Tianjin FAW Xiali, FAW Hainan Mazda and FAW Car, the second-, fourth-, tenth-, and thirteenth-ranking entities in Table 1.3, respectively, were either wholly owned subsidiaries or under the control of FAW, China's oldest automobile enterprise. Dongfeng PSA, Dongfeng Nissan, and Dongfeng Yueda Kia (listed fifth, ninth and fourteenth in Table 1.3) belong to DMC, another top carmaker in China previously known as the Second Auto Works (SAW). The "Big Three" best reveal the joint-venture model in the Chinese automobile sector: all have followed the technological strategy of establishing a joint venture with a foreign partner and producing foreign model cars to compete for a domestic market share. Shanghai VW was the first automobile joint venture in China under SAIC, and Shanghai GM was the second joint venture of SAIC. After FAW's first joint venture of FAW-VW, FAW and Mazda cooperated on FAW Car and FAW Hainan Mazda. Tianjin FAW Xiali is part of a broader cooperation between FAW and Toyota. Dongfeng PSA,

Dongfeng Nissan, and Dongfeng Yueda Kia are joint ventures established by DMC, partnered with PSA, Nissan, and Kia respectively. The only difference between the "Big Three" is that FAW and DMC are directly owned by the central government as "central enterprises" (Model 1), while SAIC belongs to the Shanghai municipal government (Model 2).

After the "Big Three", the second-tier manufacturers in Fig. 1.2 (with sales ranging from 200,000 to 500,000 units) generally correspond to the remaining corporations in Table 1.3. Changan Auto established the joint venture Changan Suzuki, Guangzhou Automobile Industry Group Corporation (GAIG) holds Guangzhou Honda, and Beijing Hyundai is a joint venture of Beijing Automobile Industry Holding Corporation (BAIC). Apart from their smaller sales, these three carmakers appear very similar to the "Big Three", being owned by either the central or local governments. However, two key outliers exist in Chery Auto and Geely Auto, respectively ranked eighth and twelfth in Table 1.3. Both are independent carmakers without joint ventures. Chery Auto is a corporation established by the local government of Anhui Province in the late 1990s. In 2003, Chery was the champion national brand of the domestic market. It was followed by Geely Auto, which not only produces cars independently, but also stands alone as a privately owned company. Chery and Geely thus became the leading corporations for Model 3 and Model 4. Two additional carmakers, Brilliance Auto and BYD Auto, are not listed in Table 1.3 but are included in Fig. 1.2 due to their rapid growth in the early 2000s and great potential to gain a further market share in the near future. Brilliance Auto belongs to the local government of Liaoning Province and has produced national cars (Model 3), while BYD Auto is a private automaker of national cars (Model 4).

Several caveats are needed when considering the pattern depicted in Fig. 1.2. First, this illustration is not a complete picture of the Chinese automobile sector, as numerous small-scale carmakers are omitted. By the start of the economic reform, China had over 100 assembly enterprises. Beyond the major automakers as listed, these small-scale assemblers account for a relatively small proportion of the market share, usually below 20% combined, and in recent years many have been absorbed by the major automakers. Among the unlisted automakers, Great Wall Motors and Anhui Jianghuai Automobile Co. (JAC) deserve a mention due to their current strong developmental progress. However, since both only recently began the transition from producing commercial vehicles to passenger cars, neither is counted in Fig. 1.2.

Second, the categorization based on ownership structure and technological strategy is intended to summarize the major attributes of each enterprise, and should not be understood in an absolute sense. The ownership structure of a modern corporation is always complicated. It is incorrect to describe FAW as wholly owned by the central government, since some of its assets are joint ventures with foreign corporations and some have gone public. Nonetheless it remains appropriate to consider FAW as centrally owned because of the dominance of national ownership within the enterprise. The same logic applies when judging the technological strategy of a firm. Among the listed assemblers, there are automakers simultaneously producing both national cars and foreign models. Taking FAW as an example again, although its main products are foreign models produced through joint ventures, FAW is also the manufacturer of the well-known national brand car, Hongqi, which was manufactured throughout the planned economy. FAW is nevertheless classified under the joint-venture production model, because of the dominance of foreign models in its overall production. The technological strategy as described in the classification thus may simplify the real operation of these enterprises, by focusing on their business structure in the early 2000s. This issue is discussed in the concluding chapter of this book.

Despite these limitations, the snapshot provided in Fig. 1.2 offers a reliable starting point for understanding the main production arrangements employed by Chinese automakers in the early 2000s. Although the sales data above relate only to 2003, the sales rankings of these automakers have remained quite stable, usually in the order of SAIC, DMC, FAW, BAIC, Changan Auto, GAIG, Brilliance Auto, Great Wall Motors, Chery Auto, Geely Auto, and JAC. The "Big Three" maintain the top spots, as can be seen, as do the respective second-tier group, although Brilliance Auto and Great Wall Motors have recently made some gains in the rankings. The production models as practiced by these automakers have also persisted, but several more joint ventures have been formed, such as Guangzhou Honda under GAIG and Changan Ford Mazda under Changan Auto. Given that the major players in the Chinese automobile market have mostly remained unchanged over the 2000s, the four production models observed should be seen not as a temporary phenomenon in the market, but as a pattern deserving investigation.

This book seeks to offer a detailed examination of the origins of the diverse production models among Chinese car manufacturers. As China's automotive production boom has gathered pace, carmakers have evolved and adapted their production arrangements according to prevailing conditions, while responding to the macro changes set in motion by the market reform and globalization. Keeping the caveats above in mind, the existence of the four production models among Chinese automakers raises puzzling questions. During the market transition, as private automobile corporations were emerging, why were the biggest assemblers still under government control? As numerous major international automakers entered the Chinese market via joint ventures, why did some domestic firms continue to make national-brand cars? These questions challenge straightforward expectations of market transition and globalization. From another perspective, clustering of automakers along the two dimensions of ownership structure and technological strategy is also intriguing. For example, enterprises owned by the central government all set up joint ventures, while national carmakers are all private companies. Interestingly, this pattern does not apply to enterprises owned by local governments. When market share is taken into consideration, the larger players, surprisingly, are in the hands of the state, pursuing the introduction of foreign models with overseas partners, while manufacturers of national cars tend to be small-sized private companies. Why would large, state-owned enterprises avoid producing national cars, leaving this duty to relatively tiny (at least in their initial stages) private enterprises? This pattern seems to go against all expectations, given the fact that Chinese state-owned enterprises had long been geared towards serving national interests in the era of the planned economy. This book aims to decode these riddles by exploring what has contributed to the variations in the four production models among Chinese automakers.

Related Literature

To address the research question above, this section reviews related theories and perspectives that may shed light on the recent development of the Chinese automobile sector. In keeping with the central themes set out above, we begin by examining literature on market transition and globalization, the two broad forces of socioeconomic transformation in China. The section then provides a concise review of theories on the technological upgrading of enterprises, focusing on three key contextual factors: firm size, affiliation with global value chains and the role of the developmental state. Lastly, the section examines specific studies on the Chinese automobile sector.

Economic Organization in Market Transition

The transformation of organizations in market transition is the subject of significant debate in the literature. The seminal work of Kornai (1979, 1986, 1992) identifies soft budget constraints as an intractable problem in the socialist economy. The fact that enterprises are often bailed out by the state regardless of operational losses tends to make shortages and lack of innovation unavoidable. Pursuing this line of thought, neoclassical economists prescribe "shock therapy" to accomplish market transition (e.g. Sachs 1989), tending to assume a linear transitional process in which market mechanisms systematically replace the old command system (e.g. Blanchard et al. 1991; Sachs 1993). Privatization of the previously state-owned assets and production organizations is considered fundamental to this process, in that "privatization is not just one of many items on the economic program. It is the transformation itself." (see Estrin 2002: 107). In this vein, the International Monetary Fund (IMF) recommended prioritizing privatization in the reform of post-communist Europe, to populate "the vacuum left by the collapse of central planning and administrative control" (Roaf et al. 2014: 12).

China's market transition, however, was not accompanied by rapid and large-scale privatization, but instead gave rise to a wide variety of property rights arrangements within organizations. Pro-market scholars consider such plurality of ownership forms represents organizational adaptation in a gradualist reform. Instead of a movement towards privatization, as in the transition societies of Eastern and Central Europe, in China a large number of collective enterprises emerged at the outset of the reform, holding opaque and vaguely defined property rights. Some scholars consider these enterprises as an organizational response to an immature market environment, in which the necessary institutions were yet to be set up (see Naughton 1994a, b). A systematic account is provided by Nee (1992), who argues that in a partially reformed institutional environment, hybrid property rights have advantages in terms of transaction costs, and the varied economic organizations that arose in China's market transition resulted from the fact that "deep structures of the reform regime are likely to reproduce important features of the state socialist redistributive economy" (Nee 1992: 22). For Nee, organizational diversity was only a temporary response, since the market reform process would improve the institutional environment so that market-based ownership structures would eventually take the lead.

Instead of emphasizing organizational adaptation in a gradual market transition, other researchers focus on the role of local governments in shaping organizational forms during the transition. For them, decentralization in China provided degrees of fiscal autonomy at local levels, and shaped incentives for local officials to get involved in economic development, creating "local state corporatism" (Oi 1992). Such a corporatist scheme combined "the entrepreneurial and governmental roles of local governments" (Oi 1996: 138) and continually incentivized local governments to extract revenue from the new market environment. By differentiating economic organizations along the socialist hierarchy, Walder (1995) further argues that by comparison with the central and provincial levels, soft budget constraints were significantly reduced at lower levels, leading local governments at these levels to become pioneers of industrial firms. This institutional arrangement is feasible in China given the fact that the Chinese planned economy had long been governed by multiple branches and regions, and was thus much decentralized compared with other socialist countries such as the Soviet Union (Qian and Xu 1993). According to this line of thought, enterprises owned, managed, and monitored by active local states were an effective alternative to large-scale privatization in driving rapid economic growth in China.

These three perspectives lead to quite different understandings about the direction and nature of China's market transition, as well as having diverging implications regarding the heterogeneous property forms in the Chinese automobile industry. The emergence of private carmakers in China matches the predictions of neoliberal thinkers, but the continued dominance of state manufacturers cautions against perceptions of a spontaneous or linear market transition within the automobile sector. Multiple ownership structures among Chinese automakers is in some ways more in line with arguments emphasizing organizational adaptation amidst partial reform, though mechanisms specific to the automobile sector remain hidden from view. With this in mind, Nee's emphasis on the legacy of the state socialist redistributive economy could be a key direction to explore in seeking to understand the evolution of organizations in the automobile sector. Local state corporatism is also a valuable tool to understand the large number of automobile enterprises owned by local states, though this alone cannot provide a complete account of the diverse trajectories that the Chinese automobile industry has experienced.

The Role of the State in Globalization

The role of the state is argued to be "at the heart of the globalization debate" (Robinson 2001: 161), the literature on which can be grouped into three major schools (Held et al. 1999; Guillén 2001). The first school maintains that the declining role of states is an unavoidable consequence of global market forces. Domestic corporations that are superseded by transnational corporations, which bypass the territorial borders of nation states, are losing their economic autonomy (Strange 1996). Capital portfolios and FDI are beyond the control of any single government, and states have been induced to abandon mercantilist regulation of financial markets (Cerny 1994). International bodies such as the WTO and IMF have also steadily replaced functions originally fulfilled by states, which have had to relinquish their role in governing national economies (Ruggie 1995).

Scholars of the second school view this globalist perspective as greatly exaggerated, arguing that states have upheld their role in the face of globalization. This position contends that most sales, assets, and technological innovations of transnational corporations remain bounded within home territories governed by states (Hirst and Thompson 1999), while most investments within a nation state still derive from domestic savings, rather than from foreign investments, with global stock markets remaining "far from fully integrated" (Wade 1996: 86). Equally, the state can bear the consequences of globalization because of its importance to national competitiveness. National innovation systems are seen as embedded within historically formed institutional arrangements, of which the state is a critical component (Hall and Soskice 2011).

A third line of thought is critical of both schools for overlooking fundamental aspects of the relationship between the state and globalization (Perraton et al. 1997). Specifically, scholars of this group tend to acknowledge the substantial impacts of globalization, albeit treating globalization as a transformative process in which the state redefines, rather than abolishing, its role in economic governance (Panitch 1995). From this position, the global economy assigns duties to the state that no alternative entity could perform, such as protecting the contract and property rights of global capital via legal systems, and maintaining stability and security in the trade and investment regimes of the global economy (Sassen 1996). The state also acts as a "transmission belt", helping the domestic economy to survive the globalized environment (Cox 1987). More importantly, new opportunities arise from globalization that allow states to assume a role in strengthening national competitiveness in the global market. States that harness industrial policies to foster close partnerships with the business community may find themselves at a great advantages in global competition. In this sense, globalization has not eroded the role of the state, but rather enhanced it (Weiss 1998).

These differing perspectives on the role of states in the face of globalization help to shed light on important aspects of our research subject, particularly in terms of how foreign joint ventures have prevailed in the Chinese automobile sector. According to the globalist view, the substantial investment by foreign corporations is certainly a critical factor for the development of the Chinese automobile sector. Yet, their numerous joint ventures with China's main automakers, and the strong and stoical presence of state actors as major shareholders in these ventures, indicate that the Chinese state has little intention of retreating under the expansion of global market forces. This scenario contradicts straightforward globalist predictions, but supports the argument for resilience on the part of the state. However, the latter offers limited insights on how the state responds to globalization within specific sectors. In this regard, the transformative view of the state in the context of globalization is perhaps a more appropriate analytical perspective, given that joint ventures in the Chinese automobile sector are strategically regulated by the state through industrial policies.

Technological Upgrading of Enterprises

Moving on from the macro processes of market transition and globalization explored above, we now turn to theoretical frames of technological innovation and upgrading among enterprises, especially in the developing countries. Three key literatures are of particular relevance here: the work of Joseph Schumpeter; theories on global value chains; and the developmental state.

The Schumpeterian Hypothesis

In his classic interrogation of capitalist and socialist economic systems, *Capitalism, Socialism and Democracy*, Schumpeter (1942) contended that large corporations in imperfect competitive markets were better at promoting technological progress. In a sharp challenge to antitrust enforcement, the Schumpeterian Hypothesis suggests that "a market structure involving large firms with a considerable degree of market power is the price that society must pay for rapid technological progress" (Nelson and Winter 1982: 278). As a central topic in studies of industrial organizations, empirical work using the Schumpeterian Hypothesis generally adopts the two independent variables of firm size and market concentration. Multiple mechanisms are proposed to establish the advantages of large firms and market concentration with regard to innovation: large firms can absorb greater R&D costs; have better external financial resources; can identify unseen innovation with diversification; enjoy economies of scale and scope in R&D; and spread risk by conducting multiple projects at the same time. Meanwhile, market concentration facilitates innovation through the profits derived from monopoly market positions and helps extract revenues from innovation (Symeonidis 1996).

Empirical applications of the Schumpeterian Hypothesis are far from conclusive, however (Cohen and Levin 1989; Symeonidis 1996). Some studies suggest the positive impacts of firm size on innovation may decline after a certain threshold (e.g. Scherer 1965), while others have found no disproportionate innovative performance among the largest firms (e.g. Link 1980). Efforts to support or refine the original model advanced by Schumpeter persist nonetheless. For example, Nicholas (2003) revealed a positive association between the market power of firms and their innovations in American industries in the early twentieth century, while Aghion et al. (2005) developed an inverted U model showing that competition–innovation associations vary at different levels of competition. The advantages of large corporations are also acknowledged in other economics literature, such as studies of economies of scale and scope (e.g. Chandler 1990) and international trade (Vernon 1971; Hymer 1976).

Scholars of the Schumpeterian Hypothesis pay much attention to variations in patterns of innovation across industries, since the relationship between firm size and innovative signature may differ by industrial structure (e.g. Acs and Audretsch 1987). That is, in sectors characterized by easy entry, innovations by small firms may prevail; whereas in sectors in which it is harder for new firms to establish themselves, larger firms are more likely to innovate (Winter 1984; Koeller 1995). The automobile sector exemplifies the latter, since high capital and technological intensity creates exceptionally high entry conditions for new firms. The Schumpeterian Hypothesis is thus a good match for the reality of the world automotive industry where a few major global automakers are largely responsible for the vast majority of innovation and technological progress.

That said, the Schumpeterian Hypothesis nevertheless seems to a great extent to operate in reverse in the case of the Chinese automobile industry. As observed, the current "Big Three" in China—FAW, SAIC Group and DMC—all concentrate on production of foreign models, while it is the relative minnows like Chery and Geely which have aggressively developed national-brand cars and sought an independent R&D capability. By introducing sophisticated foreign technology, the "Big Three" should, intuitively, be technologically more advanced than these national-brand manufacturers, yet R&D on the independent models as practiced by these small firms is outwardly more innovative. The binary contrast seen in China's car industry appears all the more intriguing in reference to the Schumpeterian Hypothesis, since this originally set out to emphasize "the qualitative differences between the innovative activities of small, entrepreneurial enterprises and those of large, modem corporations with formal R&D laboratories", rather than a continuous relationship between firm size and innovation as assumed by most subsequent research (Cohen and Levin 1989: 1067). In this sense, the bifurcation of technological strategies observed among Chinese carmakers might be considered to provide a straightforward negative case in relation to the Schumpeterian Hypothesis.

Global Value Chains

Literature on the global economy has undergone a major change of direction over the past few decades. After the Second World War, dependency theory became a key arena of criticism for the manner in which international trade continually reinforced patterns of manufactured goods originating in industrialized countries while drawing raw materials from "underdeveloped" countries, as they were then termed (e.g. dos Santos 1970; Cardoso and Faletto 1979). This core–periphery relationship was more deeply engaged in the later "world system" perspective developed by Wallerstein and his contemporaries (1974, 1980, 1989). World-system theory depicts the global economy as a hierarchical structure with unequal labor division, in which the core, namely industrialized countries, exploits the periphery, or underdeveloped countries.

These views are challenged by more recent changes in the organization of the global economy, in which connections between the core and periphery have moved significantly beyond international trade. Put another way, trade-based "internationalization" has evolved to become "globalization", in which economic activities have become increasingly functionally integrated across countries (Dicken 2003; McMichael 2003). In the face of this new stage in the world economy, new forms of production have been conceptualized as "value chains" (Kogut 1984; Porter 1985), "global commodity chains" (Gereffi and Korzeniewicz 1994) and later "global value chains" (Kaplinsky 2000; Gereffi and Kaplinsky 2001), to develop analytical frameworks in relation to the global inter-firm coordination of design, production, and marketing. A key focus in this literature is the governance structure of value chains, in which buyer-driven and producer-driven chains are first distinguished (Gereffi and Korzeniewicz 1994), and then further conceptualized into more refined categorizations (Gereffi et al. 2005). A noteworthy parallel perspective is that of global production networks, which has developed largely at the same time, and proposes moving beyond a firm-centered view to build an actor-based causal framework for globally integrated modes of production (see Dicken et al. 2001; Coe et al. 2008; Yeung and Coe 2015).

As a central topic in global value-chain analysis, governance is arguably associated with industrial upgrading, which refers to the repositioning of a firm, region or country from a lower position to a higher position in the chain. Gereffi argues that "participation in global commodity chains is a necessary step for industrial upgrading because it puts firms and economies on potentially dynamic learning curves" (1999: 39). In other words, the information flows and learning potential arising from affiliation with transnational corporations present important upgrading opportunities for enterprises in developing countries. The apparel industry is often used as a representative case for an upgrading trajectory from simple assembly to original equipment manufacture (OEM) and original brand manufacture (OBM) (Gereffi 1999; Bair and Gereffi 2001; Gereffi and Memedovic 2003). Nevertheless, it is recognized that such upgrading does not occur automatically. Although production technology can progress at various stages in the value chain under the guidance of lead firms, to take the further step of changing the role of an enterprise within a value chain may be against the interests of those lead firms, and it is often the case that firms at the upstream end of such chains are locked into their position (Humphrey and Schmitz 2002; Giuliani et al. 2005). In this situation, global value chains may "create barriers to learning and drive uneven development over time, even as they trigger rapid industrial upgrading" (Gereffi 2015: 20). Such challenges to functional upgrading within global value chains have attracted significant attention in this field.

The global value-chain framework has been applied empirically to the automobile industry. In the post-NAFTA development of Mexico's automobile sector, this framework reveals how large automotive exports were driven by global value chains led by foreign corporations (see Gereffi 2015). However, the nature of the Chinese automobile sector runs counter to a simplified application of the global value-chain approach, instead echoing recent scholarly reflections that linkages with foreign corporations may in fact block rather than facilitate functional upgrading of firms in developing countries, though their production technology may improve in the process. As observed, partnership with transnational corporations does not promote national brands in the Chinese automobile sector, in stark contrast to the struggles of independent national carmakers lacking foreign partners. Sturgeon et al. (2008) introduced three further elements to the global value-chain perspective in their analysis of the global automobile industry: firm-level chain governance, power, and institutions. This extended framework accounts more effectively for important characteristics in the global automobile industry, such as how local content requirements have helped shape regionalization of global automobile production. Likewise, it is necessary to introduce new theoretical elements if global value chains are to explain production variations among Chinese automakers—especially in terms of the factors behind the upgrading of those Chinese carmakers that operate independently from global producers.

The Developmental State

Whether and how states should intervene in markets is an age-old debate. Neoclassical economists consider that state involvement in markets generates inefficiency, as well as an inherent rigidity that distorts market mechanisms. This position sees the state as incompetent to handle problems of information that arise in economic administration, and state interventions may also lead to corruption on the part of rent-seeking bureaucrats (see Evans 1997; Chang 2003). For these reasons, the state should maintain a "night watchman" role with respect to markets and minimize interference in their operation (Buchanan et al. 1980). For critics of the neoliberal ideology that these perspectives have led to, market governance has become exaggerated as an inevitable and self-regulating force (Gore 2000; Steger 2005). Neoliberalism has also lost its standing in some quarters as the world economy has gained little in terms of growth, employment, and especially equality, under its direction (Cohen and Centeno 2006). The unregulated market alone cannot achieve socially optimal levels of economic welfare, and may yet lead to a serious crisis of capitalism itself (Sassen 1996; Panic 2007). Without state intervention, business cycles could be exacerbated, while some countries may risk plummeting to the bottom in the competition to attract foreign investments (Yeung 2000).

Scholars of developmental state theory further propose a proactive state as the central agent of development, driving industrial upgrading and economic growth in developing countries. In the 1980s and 1990s, the miraculous economic success of the East Asian "tigers" inspired a slew of studies underscoring the importance of the state in the development process (e.g. Johnson 1982 on Japan; Amsden 1989 on South Korea; Wade 1990 on Taiwan). A developmental state, as conceptualized based on these countries, commonly has autonomous economic goals regarding industrial upgrading, and seeks to enact these via a cohort of efficient technocrats armed with effective policy protocols to coordinate and mobilize domestic business. Although developmental state theory is axiomatically state centered, it does not simply deny the role of the market, but rather proposes state interventions as an alternative when market mechanisms are weak (Aoki et al. 1998).

Arguments in favor of developmental states were challenged in the wake of the Asian financial crisis in the late 1990s, though it would be hard to consider the developmental state extinct, particularly in the East Asian region. Where the developmental state might be considered to some degree to have been replaced by a neoliberal state, this is criticized as exaggeration by some scholars (Walter 2006). In some East and Southeast Asian countries, state incumbents fought hard to avoid being overrun by the intrusion of neoliberal policy prescriptions after the financial crisis. Recent studies on these countries have found that the developmental role of state agencies, far from waning, has evolved or even strengthened since the crisis (Dent 2003; Pereira 2008). Notably, the role of the state in development is once again gaining attention in mainstream economics. As recently proposed in new structural economics (see Lin 2012), to trigger transformation across different developmental stages, states could provide information about new industries needed, promote coordination of infrastructure investments, compensate externalities among the first movers in upgrading, and otherwise attract FDI.

In studies of the developmental state, the automobile industry has been cited as a representative case for how such interventions in an economic sector could accelerate the technological upgrading of national assemblers (see Amsden 1989). However, if this state-centered approach is applied to the Chinese automobile industry, the expectation that enterprises with state support make better progress technologically is not completely accurate. Although enterprises directly under state guidance such as FAW, DMC, and SAIC (the "Big Three") have seen technological improvements by manufacturing advanced models introduced by foreign partners, this again is not at all comparable to the experience of national carmakers such as Chery Auto and Geely Auto, who conducted independent R&D with national brands. Far from being nurtured by developmental state interventions, China's fledgling national carmakers were frequently not officially endorsed, and thus technically inhibited in their early stages of growth.

The limitation of developmental state theory in explaining variations among Chinese automakers is shared by the Schumpeterian Hypothesis and the global value-chain framework as discussed above. If the straightforward predictions of these three approaches are considered together, large enterprises with a close affiliation with advanced foreign partners, coupled with strong state support, should progress better in technological upgrading. In the reality of today's China, this scenario is only partially accurate: large state-owned assemblers in China have indeed acquired more advanced production techniques, better production management, and the ability to manufacture prominent car models internationally through joint ventures with foreign partners. On the other hand, these firms lag behind small-scale domestic assemblers with no foreign partners and negligible state support in terms of developing self-reliant R&D of complete cars, a more innovative form of technology upgrading. The existence of these proactive national carmakers, and how they have been able to achieve this, are thus key aspects of the puzzle central to this book.

Studies of the Chinese Automobile Industry

Prominent existing scholarship on the Chinese automobile sector includes Mann's study (1997) of Beijing Jeep, the earliest car joint venture in China; Gallagher's book (2006) on the energy and environmental implications of the Chinese automobile sector; and Chin's recent study (2010) on how China has governed foreign automobile companies through industrial policies. Among these, two studies stand out as most relevant to the current book, based on a shared primary concern to understand how the organization of automobile production has changed in post-reform China. Eric Harwit's book (1995), *China's Automobile Industry: Policies, Problems, and Prospects*, investigated major foreign-invested joint ventures for passenger cars in China from the 1980s to the early 1990s, with the goal of understanding behind-the-scenes policy-making processes. This study captured the central theme of China's automobile industry in the early stages of the market reform, namely how foreign automakers could get into China. A series of central policy decisions to bring in global carmakers in the 1980s and 1990s was a milestone for China to develop large-scale, market-oriented production of passenger cars. Harwit provides detailed descriptions and insightful accounts of this historical process. The theoretical framework in the book emphasizes the motivations and ideas of individual and organizational actors relating to the passenger-car projects, and examines subsequent policy-making processes characterized by constant bargaining and misconceptions. Based on four empirical cases, Beijing Jeep, Shanghai-VW, Guangzhou-Peugeot and Panda Motors, this book reveals the complicated political and economic interactions between foreign corporations, the central government and its ministries, and local states in passenger-car projects during this period.

The second key work is *Changing Lanes in China: Foreign Direct Investment, Local Governments and Auto Sector Development* by Eric Thun (2006). This book examines two periods in the development of the Chinese automobile industry: the 1980s to the late 1990s and the late 1990s to the early 2000s, with a focus on how local automobile sectors developed supply networks. Seeing local opportunities to develop a modern automobile sector emerging from the establishment of joint ventures with foreign manufacturers, Thun sought explanations for the regional variations in performance in the nurturing of supply networks. The local institution, or locality, is proposed by Thun for explanatory purposes, which is then operationalized as the local intra-firm relationship (hierarchical versus market) and local bureaucratic structure (unified versus fragmented). In this framework, the role of the local state is persuasively underscored as a key factor in understanding developmental variations across regions. Three types of locality are distinguished among Thun's five empirical cases: Shanghai as a local developmental state; Beijing and Guangzhou as laissez-faire local states; and Changchun and Wuhan as firm-dominated localities. This typology is then applied to explain the successes and failures of local supply networks. The local developmental state of Shanghai is identified as the most successful model, though Thun argues that this model faced mounting challenges as the Chinese automobile market gradually matured.

The development characteristics of the Chinese automobile sector have varied dramatically by historical period, and the present study aims to offer a timely follow-up investigation of this dynamic industry, moving forward from the two previous studies. Both Harwit's and Thun's work focused on issues related to the joint venture, which has been the typical organizational form for manufacturing cars in China during the past three decades. Briefly stated, Harwit concentrated mostly on how joint ventures were established in the 1980s and 1990s, whereas Thun was more concerned about how joint ventures could be successful in the market through developing local supply networks after the 1990s. Not until very recently have new types of player emerged in the Chinese car market: independent carmakers who do not rely on Western partners, and privately owned firms without state involvement. This book captures the significant changes that have allowed the existence of and been brought about by these new players, and aims to explore how their production models (Models 3 and 4) are possible alongside the more traditional models featuring state-led establishment of joint ventures (Models 1 and 2). By bringing these new players under scrutiny, this book covers a wider spectrum of production varieties within the Chinese automobile sector than previously analyzed. In this sense, it is not merely a continuation of the two previous studies; rather it provides an updated discussion accounting for the transformation of the Chinese automobile industry that has since occurred.

An Institutional Framework

While the literature considered above offers useful approaches to the central concerns of this book, ultimately it explains little about the origins of production variation among Chinese carmakers. Current macro-structural changes in the Chinese automobile sector, namely endogenous marketization and exogenous globalization, are crucial to understanding diversity of ownership structures and the emergence of joint ventures in the sector, but they are not sufficient to reveal how each specific production model was adopted by the automakers. On the other hand, literature on technological upgrading highlights organizational attributes such as firm size, international affiliation and availability of state support, all of which implicitly assume that enterprises are autonomous in a stable market economy. In any analysis of Chinese car enterprises, however, such an assumption would miss the key contextual point that these enterprises were not given, but born out of the sweeping institutional transformations of market reform. China's auto manufacturers are thus better understood as multi-actor projects forged in the furnace of an emerging and unpredictable market, rather than independent actors in a given, stable market. To address these limitations, this book proposes an institutional explanation for the varying developmental trajectories of Chinese carmakers in the three decades since the instigation of the market reform.

The proposed analytical framework of this book will examine China's carmakers since the 1980s as local industrial projects that involve multiple stakeholders including the central government, local governments, and local enterprises and business actors. Due to the economic significance of the sector, these stakeholders have developed strong interests in automobile manufacturing, but their political power, rationales, strategies, and actions pertaining to these projects have been heavily shaped by institutional forces, which in turn have driven the car projects along different developmental tracks and led to the resulting variety of production models. The relevant major institutional forces can be grouped under three headings: the national positioning of the local economy within the previous central planning regime; the local political structure during the market reform; and the configuration of local developmental ideas during the market reform. I argue that the pre-transition national positioning of a given local economy created path dependencies that shaped the local political structure and configuration of developmental attitudes towards car projects during the reform. The local political structure enabled and regulated relevant stakeholders' distinct interests in the project, whereas the configuration of developmental ideas stimulated and nurtured the rationales and strategies adopted by these stakeholders. From this perspective, this book considers the ownership structure and technological strategy of Chinese car assemblers since the 1980s as outcomes of local socio-political processes, characterized by power-differentiated interactions among related actors with differing developmental ideas. This analytical framework is illustrated in Fig. 1.3.

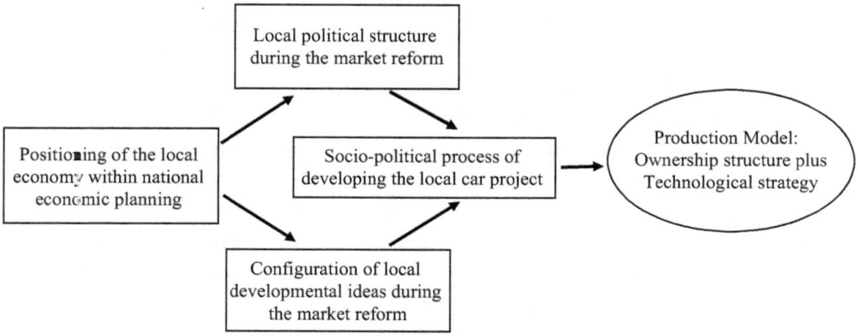

Fig. 1.3 Analytical framework

This proposed framework incorporates and expands both Harwit's and Thun's theoretical perspectives. Developmental ideas about introducing foreign-invested joint ventures, the central element of Harwit's analysis of the policy-making process regarding joint ventures in the Chinese automobile industry, are highlighted here as an important cognitive resource for some production models. However, the proposed framework also depicts scenarios beyond the central official ideology and related policy discussions on joint ventures highlighted by Harwit. That is, the developmental ideas of different levels of actors about car enterprises could also be contradictory rather than consistently top down. For example, it is possible to analyze the developmental rationale of independent manufacturers alongside, and often in contradiction to, national industrial policy. Although Thun's perspective on local institutions is largely adopted in the current analysis, this study adds to Thun's work in two ways. First, different sets of institutional factors are brought into the analytical frame by considering the historical positioning of the local economy in the communist planning regime, together with the associated local political structure and developmental ideas during and after market reform. Second, in contrast to Thun's framework, which highlights the implications of central regulations for the locality before focusing on the role of the local state in developing supply networks, this study delves into historically rooted and variegated coordination and conflict between the central and local state, as well as between the central state and local enterprises. In integrating political and cognitive factors, the proposed framework resembles the well-established political-cultural approach advanced by Neil Fligstein (1996, 2003), which argues that market actors constantly interact with the state to seek social solutions for competition. However, the unique context of the Chinese automobile sector drives the need for the framework proposed here to differ from Fligstein's theory. In this regard, the production models of Chinese automakers are seen as deeply embedded in the market transition, rather than evolving from the interaction of market actors as assumed by a generalized setting.

In sum, based on theoretical perspectives from established literature, the analytical framework of this book traces diversity in the organization of car production in China back to local economic positioning within the previous planned-economy system, its enduring effects on the local political structure and developmental ideas in the market reform, and the resulting socio-political processes that have shaped the development of car projects. Each of these analytical components is expanded on below.

National economic positioning refers to the designation of economic functions for the national economy applied by the central government to a specific region. Before the market reform, one of the central government's major strategies for planning and managing the national economy of China was to designate and adjust the economic functions of the country's regions. This practice is best illustrated by China's top-level economic blueprints during the command-economy era: the five-year plans. China's pre-reform five-year plans drove three successive waves of regional economic positioning. The first five-year plan (1953–1957) concentrated on the heavy industries, in which the Northeast was chosen as the main target of investment. This region was prioritized based on its proximity to the border with the Soviet Union, the main source of economic assistance in the initial phase of the regime, as well as the fact that it contained the bulk of the industrial infrastructure constructed during the Japanese colonization. From the late 1950s, the second five-year plan (1958–1962) aimed to develop seven economic zones, each with an independent industrial system: the Northeast, North, East, Southwest, Northwest, Central, and South zones. This ambitious initiative was one aspect of the ill-fated Great Leap Forward movement, and ultimately proved unsuccessful. Starting from the late 1960s, the third (1966–1970) and fourth (1971–1975) five-year plans were proposed under the Third Front Movement, which aimed to develop military-related industries in the western regions (especially the Southwest), to strengthen national defense against potential war threats arising from Sino-Soviet tensions. The centrally directed regional economic positioning did not simply amount to words in the pages of official documents, but had significant impacts in terms of the distribution of the national budget, training, and transfer of technicians and labor, appointment of directors and managers, and construction of infrastructure and machinery. Consequently, in the period of the planned economy positioning activities under the five-year plans, in which automobile manufacturing was a priority, determined the geographical distribution of major industrial sectors for many years thereafter. Two of the major national auto projects were FAW, located in the northeast industrial heartland, and Second Auto Works (SAW) later DMC, located in the west, whose fortunes were respectively set in motion by the first and third waves of national economic positioning.

The major role of national economic positioning in the planned economy also had a lasting influence on China's market transition, shaping the developmental trajectories of car-manufacturing projects from the 1980s.

The path dependency according to which the former economic planning influences the new market is broadly referred to as the "lock-in" effect of historical legacy on later trajectories. Economists first developed this concept to explain the existence of inefficient technological standards, proposing that the increasing returns could encourage the inefficient to be resilient against alternatives (David 1985; Arthur 1989). Neo-intuitionalist economics later introduced bounded rationality and high transaction costs to explain such institutional inertia (North 1990; Williamson 1993). Besides economics perspectives, Thelen (1999) has distinguished additional political mechanisms of (1) shared cognitive frames that sustain established institutions and (2) "positive feedbacks" in politics. Other classifications to enhance the explanatory capabilities of path dependency have been developed, for example, by Mahoney (2000) who proposed four overall types of institutional path dependency (utilitarian, functional, power and legitimation); and by Beyer (2010) who in a recent review observed seven mechanisms: increasing returns, irreversibility, functionality, complementarity, power, legitimacy, and conformity.

The perspective of path dependency has been applied to varieties of economic organization in different institutional contexts. Hamilton and Biggart (1988), for instance, attributed the differences between economic organization in the three East Asian economies of Japan, Korea, and Taiwan to historical state–enterprise relationships. Biggart and Guillén (1999) compared automotive production in Korea, Taiwan, Spain, and Argentina, concluding that variations between these countries were largely due to specific socio-organizational legacies, which created unique capacities and pathways for each nation to compete in the global automobile market.

The effect of path dependency extending from national economic positioning in the central planning era, as highlighted in this analysis, is largely down to the distinctive nature of China's reform. The reform in China has in reality been tightly planned and controlled by a strong central government with its underlying ideology and political system remaining largely intact (Liew 1995). When new economic institutions were introduced in the reform, previous institutions persisted to some extent, and continued to exert their influence. It is in this sense that, as Nee (1989) observes, the regime's fundamental structures and socialist redistributive mechanisms would most likely replicate themselves to some extent, even as organizational forms began to adapt to new conditions. This reflects the gradualist rationale of China's market reform, though debate over whether incremental reform is superior to the shock therapies of the Eastern European countries remains ongoing (see Benziger 1998; Woo 1999; Iwasaki and Suzuki 2016).

The effect of path dependency on the post-reform automobile industry mainly unfolded in local political structures and the configuration of local developmental ideas. In this book, the local political structure refers to the relationship between major actors in local political-economic affairs, the pivotal institutional architecture enabling relevant actors to develop car-manufacturing projects, as well as regulating their interactions. After the market reform, numerous local governments, enterprises, and executives were released to the new market as (semi-) autonomous market actors, though the central government remains the dominant force of intervention in the national economy. New automobile production initiatives involving substantial investment have significant economic implications, and thus always pertain to the major stakeholders in the local economy. The interests of these stakeholders are not necessarily consistent, however. In the emerging market environment, the central government was most concerned about how local car projects would fit in with national automobile production plans, while local governments highlighted their potential to boost local economies, and entrepreneurs obviously focused on market opportunities. The existence of differing interests created the conditions for competition among these actors, such that any local trial to manufacture passenger cars might potentially become politicized within the domain of local power hierarchies as "a political project undertaken by powerful actors" (Fligstein 1996: 657).

Importantly, these actors are not "freely-contracting individuals" but operate in "a world in which institutions give some groups or interests disproportionate access to the decision-making process" (Hall and Taylor 1996: 9). That is, the power distribution among these actors is often framed by institutional arrangements. In this case, pre-existing national economic positioning was a main determinant for local political structures in the market transition. Prior to the reform, the layout of major industrial projects across China, as shaped by national policies such as the five-year plans, led to an uneven distribution of national resources, and contributed to the varying political significance of local governments in the eyes of the central planners. That is, the function of a local economy on the national economic map was usually associated with the political clout of the local officials. In regions occupying core economic positions, such as those hosting industrial and technology bases, local officials often held higher political ranks, had better promotion prospects, and gained more attention and support from above. In contrast, officials in peripheral regions had relatively lower political influence and weak voices on the national planning agenda. During the market transition, these pre-existing relations "continue to evolve in response to changing environmental conditions and ongoing political maneuvering but in ways that are constrained by past trajectories" (Thelen 1999: 387). Faced with the new market environment, regions with existing advantages under the planned economy tended to retain central support for their own economic projects, and/or enjoyed a greater prospect of once again being trusted by central planners to carry out projects of national economic importance in the new era. On the other hand, in the regions that had been previously regarded as peripheral, local officials were much less likely to benefit from central sponsorship, and had to battle for opportunities in the emerging market. That said, for some regions which had been particularly marginalized in the previous system, an interesting reverse-feedback mechanism seemed to arise: in some instances, so weak was the legacy of the planned economy that this actually served to promote an active private economy, in which private enterprises and actors took the lead in developing the local economy. These scenarios demonstrate how the influence of historical political legacies of the pre-reform era might persist through the market transition, affecting how the central and local governments, as well as local enterprises and business actors, may become involved in a complex, capital-intensive project such as automobile manufacturing.

The configuration of local development ideas in this book mainly refers to the pattern of views among local actors on whether and how to develop automobile production, in other words the cognitive resource stimulating and nurturing the rationale and strategy adopted by these actors in establishing a car project. Despite the great potential for economic returns from such projects, they nevertheless involve huge risks. The decision to launch a car project is thus rarely based on a completely calculable procedure, and always entails a sense-making process which gives meaning to these risk-taking behaviors. Meaning is in turn always contested, as observed by Weber: "in no case does [meaning] refer to an objectively 'correct' meaning or one which is true in some metaphysical sense" (1978: 4). In this sense, despite its deeply practical concerns in terms of capital and technology, a car project can also be considered an essentially cognitive and normative discourse, in which developmental ideas permeate the local decision-making process and function as a road map for the articulation of the appropriate arrangement of actions (Dobbin 1993; Goldstein and Keohane 1993). While a car project may match the ideas of all related actors, it is also likely that cognitive factors will remain "always contested and ultimately are never settled" since "the content of ideas alters over time or in the course of conflicts and compromises" (Jacobsen 1995: 288–89, 300). This is to say that the discourse relating to a given car project may be driven by a configuration of coordinated or contested perceptions of the automobile sector among the stakeholders involved.

For each specific region, ideas on whether and how to develop the production of cars did not arise spontaneously after the market reform, but could again be traced to the previous period of the command economy. Specifically, the post-reform configuration of local developmental ideas surrounding car production largely originated in each region's former national economic positioning. From the beginning of the planned economy, China's communist leaders were strongly committed to the national imperative of industrialization. The expansion of the automobile sector was given high policy priority throughout the planned-economy period. Starting from scratch, the central leaders decided to concentrate national resources on a few key enterprises. This approach was also adopted as a major principle in developing many other sectors, though it was provisionally discarded during the Great Leap Forward. The core ideologies and objectives of the command economy nevertheless persisted during the market reform. As the market began to gain traction, the importance of the automobile sector and the necessity of a state-led effort to preserve industrial concentration became watchwords of national industrial policy. Confining automobile production within a few established enterprises was well received among the old guard of central planning. However, the severe market entry regulations that arose from this policy direction were much less popular among actors harboring ambitions to establish their own car enterprises in the new market environment. In particular, what appeared reasoned and logical for the central government was perceived as unfair among local governments, businesses, and entrepreneurs that had been depressed, stifled, and ignored under the planned economy. The central planners' official developmental ideology thus encountered allies and enemies in different regions, depending on their previous national economic positioning.

The production models of Chinese carmakers after the reform, characterized by specific ownership structures and technological strategies, arose from a context-specific socio-political process of interaction between actors with different political powers and developmental ideas in relation to each project. For a post-reform car project in China, ideas and interests among the related actors could not be considered "phenomenologically separate" (Goldstein and Keohane 1993). Actor interests with respect to a local car project, whether central government, local government or enterprise-based, were at all times subject to certain cognitive rationales, because ideas, developmental or otherwise, have the capacity to "change the perceptions a group had of its own interests" (Hall 1989: 369). On the one hand, without underlying interests, it would be hard for particular economic ideas to gain credence among related actors, and it is thus often the case that ideas relating to car projects were unitized by actors "that inevitably inflect ideas they adopt–whatever their origin–to serve their own needs" (Jacobsen 1995: 288). Thus, behind each car project was simultaneously a political and a cognitive process, in other words a socio-political process.

Though institutional factors are clearly fundamental in the context of this study, agency cannot be ignored when seeking to elucidate such socio-political processes (see DiMaggio 1988). Political and cognitive structures may operate as institutional infrastructures, enabling and guiding related actors, but not automatically steering the socio-political process towards a given end. It was the individual agency of local actors that created strategies, mobilized resources, and commenced production; and it was also the agency of these actors that constantly generated interactions, whether contesting or coordinating (Campbell 2002). As China had no experience of mass production of cars prior to the reform, a car project must always have been a daunting task, even for central policymakers. Those who committed to this sector would have to be of a persevering and innovative nature to overcome the intrinsic challenges relating to, among other things, finance, technology, and infrastructure. More importantly, they had to be flexible and strategic in dealing with other actors who might have opposing interests and concerns in the same project. This was especially true for local governments and enterprises which attempted to develop car production below the radar of the central authority, and against the tide of prevailing national industrial policy. Each of these issues forms an important aspect of understanding the socio-political processes that underpinned the establishment of car enterprises in reform-era China.

Case Selection

This book examines the validity of the theoretical arguments set out above, focusing on four cases: FAW, SAIC, Chery Auto, and Geely Auto. These four cases were purposively selected from a sampling framework comprising China's major carmakers listed in Fig. 1.2, each representing one of the four production models. While other carmakers can also be categorized under each model, the selected cases exhibited the most significant sales, were the most stable in terms of their status within the market, and the most consistent practitioners for the production model they represent. Thus, of the three automakers under Model 1 as listed in Fig. 1.2 (i.e. all centrally owned enterprises engaged in joint ventures with foreign partners), FAW stands out as the strongest case. Referred to as the "senior son" of the Chinese automobile sector, FAW was the first automobile enterprise established after the founding of communist China, and has remained firmly among the "Big Three" following the reform. The consistency of FAW's development, despite the changes in industry and management structures during the reform years, provides a window through which to observe organizational change in the context of market transition specific to the institutional characteristics of China. DMC, originally SAW, was a slightly later entrant to the sector, having been built in the 1960s as a national project, while Changan Auto was historically a military enterprise and only converted to civilian production after the reform. Of the carmakers categorized under Model 2, those owned by local governments and cooperating with foreign corporations, SAIC was selected owing to the combination of being the current largest carmaker in China and pioneer of China's first passenger-car joint venture. Other enterprises representing Model 2 include GAIG and BAIC, neither of which is comparable with SAIC in terms of scale of operation or historic role in bringing the joint-venture model to the Chinese automobile industry.

Of the two enterprises listed under Model 3, Chery Auto has most consistently practiced a production model characterized by local government ownership and independent technological upgrading. The other option, Brilliance Auto, is not a pure case for Model 3, as in addition to autonomously developed national cars, Brilliance Auto has a joint venture with BMW, thus producing foreign models as per the first two production models. Moreover, there were major legal disputes over the ownership structure of Brilliance Auto during the early 2000s. Finally, of the two private independent carmakers under Model 4, Geely Auto was selected in preference to BYD Auto owing to its longer history and larger output.

Geely has also visibly followed a more stable and complete development trajectory, having started out as a private carmaker in the late 1990s and concentrated on passenger cars until the present day. In contrast, BYD Auto was previously a successful manufacturer of batteries, and only ventured into car manufacturing in 2003 after purchasing Qingchuan Auto, a small Chinese automaker in the northwest.

To identify the geographical spread of the selected cases in relation to the rest of the industry, Fig. 1.4 maps China's thirteen major automakers in 2008. The map shows the distribution of Chinese carmakers, primarily across the country's eastern provinces. Among the four selected cases, FAW lies in the northeast, while SAIC, Geely, and Chery are clustered around the mouth of the Yangtze River, though each belongs to a separate province or municipality. Empirical data used in the book are mostly based on qualitative interviews, historical archives, government documents, and media reports collected throughout the fieldwork in these four cites.

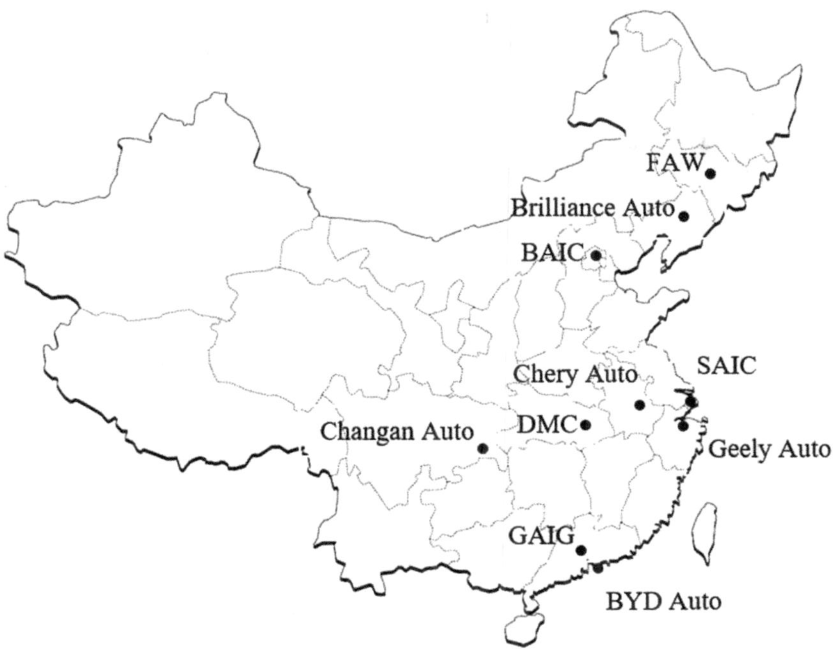

Fig. 1.4 Geographical distribution of major car enterprises in China in the early 2000s

Outline of the Book

In the empirical chapters that follow, each of the four selected cases is investigated in turn, followed by a concluding chapter that provides a systematic summary and related discussion of the theoretical implications. Each empirical chapter applies the analytical frame set out above to analyze the development of FAW, SAIC, Chery Auto, and Geely Auto, respectively. These chapters thus first review the pre-reform position assigned to the home region of each carmaker within the national economy, before considering how such historical arrangements shaped local political structures and the configuration of local developmental ideas during the reform. The chapters then detail the resulting socio-political processes that gave rise to the unique ownership structures and technological strategies of the carmakers studied. A brief case summary of key aspects in relation to the theoretical arguments of the book is provided at the end of each empirical chapter.

Chapter 2 traces the origins and development of China's oldest automobile enterprise, FAW, representing Model 1, through the era of the planned economy, during which the central government designated China's Northeast as the national heavy-industry base, making FAW responsible for providing vehicles to support industrialization. This historical legacy underpinned the central government's desire to retain control over FAW as a leading exponent of national industrial policy once the market reform started. The chapter describes how FAW became an expression of the national socio-political will in the reform, transforming from a state-owned enterprise in the planned economy to a state-owned corporation cooperating with foreign partners.

Chapter 3 examines SAIC, China's largest carmaker, representing Model 2. The origin of SAIC is to be found in Shanghai's historical status as the national economic center during the communist period, which granted local government leaders unique political leverage. Such advantages enabled Shanghai government officials to garner central support for local manufacturing, once they were committed to developing the automobile sector during the market transition. In consequence, Shanghai successfully lobbied the central authority for the then unprecedented establishment of a joint venture with a foreign corporation, a notion of great political sensitivity at the time. SAIC's unique model, in sum, was the product of a strong and active local government and the hard-won sponsorship of the central government.

Chapter 4 visits Chery Auto, representing Model 3, to illustrate how local government actors labored to develop an enterprise—partially below the radar of the central authority—for the manufacture of national cars during the market reform. In the planned economy, Chery Auto's home province of Anhui was assigned to provide raw materials to assist industrialization of more advanced regions such as Shanghai. Long neglected in national industrialization plans, local officials harbored lofty ambitions towards industrial projects such as car manufacturing once the reform started. The central authorities, however, with more than one hundred assemblers already active, were deeply reluctant to allow any new market entrant that might threaten the stability of the national automobile sector. This chapter follows the long battle between the central and local governments, in which Chery Auto struggled in secret to "hijack" official regulations for its own ends, as well as to pursue a technological path without foreign partners.

In the last empirical chapter, Geely Auto is investigated as a fully private manufacturer of national cars (Model 4). As with the previous cases, the advent of this production model was particularly associated with the unique national economic positioning of Geely's home province of Zhejiang in the pre-reform years. A barren land with a defensive frontier with Taiwan, Zhejiang had long been overlooked by the communist regime. The shallow legacy of the planned economy thus led to an active private economy once the market reform took hold. More so than Chery, Geely Auto's grassroots car-production ambitions seriously violated national regulations aimed at preserving industrial concentration. Chapter 5 details the relentless struggle of this independent automaker and its charismatic management to achieve legitimate status and autonomous technological upgrading.

The final chapter summarizes the empirical findings of the four cases with a systematic comparison and provides a concluding discussion. The validity and limitations of the proposed analytical framework are systematically examined, together with the broader theoretical implications of this study for literature on market transition and globalization.

Bibliography

Acs, Z. J., & Audretsch, D. (1987). Innovation in Large and Small Firms. *Economic Letters, 23*, 109–112.

Aghion, P., Bloom, N., Blundell, R., Griffith, R., & Howitt, P. (2005). Competition and Innovation: An Inverted-U Relationship. *Quarterly Journal of Economics, 120*(2), 701–728.

Amsden, A. (1989). *Asia's Next Giant: South Korea and Late Industrialization*. New York: Oxford University Press.

Aoki, M., Kim, H.-K., & Okuno-Fujiwara, M. (1998). *The Role of Government in East Asian Economic Development*. New York: Oxford University Press.

Arthur, W. B. (1989). Competing Technologies, Increasing Returns, and Lock-In by Historical Events. *Economic Journal, 99*, 116–131.

Bair, J., & Gereffi, G. (2001). Local Cluster in Global Chains: The Causes and Consequences of Export Dynamics in Torreon's Blue Jean Industry. *World Development, 29*(11), 1885–1903.

Bennett, D. C., & Sharpe, K. E. (1985). *Transnational Corporations Versus the State: The Political Economy of the Mexico Auto Industry*. Princeton: Princeton University Press.

Benziger, V. (1998). Can China's Gradualist Reform Strategy Be Applied in Eastern Europe? *Journal of the Asia Pacific Economy, 3*(1), 38–60.

Beyer, J. (2010). The Same or Not the Same: On the Variety of Mechanisms of Path Dependence. *International Journal of Social Sciences, 5*(1), 1–11.

Biggart, W., & Guillén, M. (1999). Developing Difference: Social Organization and the Rise of the Auto Industries of South Korea, Taiwan, Spain and Argentina. *American Sociological Review, 64*(5), 722–747.

Blanchard, D. R., Krugman, P., Layard, R., & Summers, L. (1991). *Reform in Eastern Europe*. Cambridge, MA: MIT Press.

Buchanan, J. M., Tollison, R. D., & Tullock, G. (1980). *Towards a Theory of Rent-Seeking Society*. College Station: Texas A&M University Press.

Campbell, J. L. (2002). Ideas, Politics, and Public Policy. *Annual Review of Sociology, 28*, 21–38.

Cardoso, F. H., & Faletto, E. (1979). *Dependency and Development in Latin America*. Berkeley: University of California Press.

Cerny, P. G. (1994). The Dynamics of Financial Globalization: Technology, Market Structure and Policy Response. *Policy Sciences, 27*(4), 319–342.

Chandler, A. (1990). *Scale and Scope: The Dynamics of Industrial Capitalism*. Cambridge, MA: Belknap Press.

Chang, H.-J. (2003). *Globalization, Economic Development and the Role of the State*. New York: Zed Books.

Chin, G. T. (2010). *China's Automotive Modernization: The Party-State and Multinational Corporations*. New York: Palgrave Macmillan.

China Auto News. (2005). *Chinese Automobile Industry Full Record, 1949 to 2004*. Beijing: Chinese Auto News. (in Chinese).

Coe, N. M., Dicken, P., & Hess, M. (2008). Global Production Networks: Realizing the Potential. *Journal of Economic Geography, 8*(3), 271–295.

Cohen, J. N., & Centeno, M. A. (2006). Neoliberalism and Patterns of Economic Performance, 1980–2000. *The Annals of the American Academy of Political and Social Science, 606*, 32–67.

Cohen, W. M., & Levin, R. C. (1989). Empirical Studies of Innovation and Market Structure. In R. Schmalensee & R. D. Willig (Eds.), *Handbook of Industrial Organization* (Vol. II). Amsterdam: North-Holland.

Cox, R. W. (1987). *Production, Power and World Order.* New York: Cambridge University Press.

David, P. A. (1985). Clio and the Economics of QWERTY. *American Economic Review, 75*(2), 332–337.

Dent, C. M. (2003). Taiwan's Foreign Economic Policy: The 'Liberalization Plus' Approach of an Evolving Developmental State. *Modern Asian Studies, 37*, 461–483.

Dicken, P. (2003). *Global Shift: Reshaping the Global Economic Map in the 21th Century.* New York: The Guilford Press.

Dicken, P., Kelly, P., Olds, K., & Yeung, H. W. (2001). Chains and Networks, Territories and Scales: Towards an Analytical Framework for the Global Economy. *Global Networks, 1*(2), 89–112.

DiMaggio, P. (1988). Interest and Agency in Institutional Theory. In L. Zucker (Ed.), *Institutional Patterns and Organizations: Culture and Environment* (pp. 3–12). Cambridge, MA: Ballinger.

Dobbin, F. (1993). The Social Construction of the Great Depression: Industrial Policy During the 1930s in the United States, Britain, and France. *Theory and Society, 22*(1), 1–56.

dos Santos, T. (1970). The Structure of Dependence. *American Economic Review, 60*(2), 231–236.

Estrin, S. (2002). Competition and Corporate Governance in Transition. *Journal of Economic Perspectives, 16*(1), 101–124.

Evans, P. (1997). The Eclipse of the State? Reflection on Stateness in an Era of Globalization. *World Politics, 50*(1), 62–87.

Fligstein, N. (1996). Market as Politics: A Political-Culture Approach to Market Institutions. *American Sociological Review, 61*(4), 656–673.

Fligstein, N. (2003). *The Architecture of Markets: An Economic Sociology of Twenty-First-Century Capitalist Societies.* Princeton: Princeton University Press.

Gallagher, K. S. (2006). *China Shifts Gears: Automakers, Oil, Pollution, and Development.* Cambridge, MA: MIT Press.

Gereffi, G. (1999). International Trade and Industrial Upgrading in Apparel Commodity Chain. *Journal of International Economics, 48*(1), 37–70.

Gereffi, G. (2015). Global Value Chains, Development and Emerging Economies. *Inclusive and Sustainable Industrial Development Working Paper Series (WP 18).* United Nations Industrial Development Organization.

Gereffi, G., & Kaplinsky, R. (2001). The Value of Value Chains: Spreading the Gains from Globalisation. *Special Issue of the IDS Bulletin 32.* The Institute of Development Studies.

Gereffi, G., & Korzeniewicz, M. (1994). *Commodity Chains and Global Capitalism.* Westport: Praeger.

Gereffi, G., & Memedovic, O. (2003). *The Global Apparel Value Chain: What Prospects for Upgrading by Developing Countries* (Sectoral Studies Series). Vienna: United Nations Industrial Development Organization.

Gereffi, G., Humphrey, J., & Sturgeon, T. (2005). The Governance of Global Value Chains. *Review of International Political Economy, 12*(1), 78–104.

Giuliani, E., Rabellotti, R., & Pietrobelli, C. (2005). Upgrading in Global Value Chains: Lessons from Latin American Clusters. *World Development, 33*(4), 549–573.

Goldstein, J., & Keohane, R. (1993). *Ideas and Foreign Policy: Beliefs, Institutions and Political Changes*. New York: Cornell University Press.

Gore, C. (2000). The Rise and Fall of the Washington Consensus as a Paradigm for Developing Countries. *World Development, 28*(5), 789–804.

Green, A. E. (1992). South Korea's Automobile Industry: Development and Prospects. *Asia Survey, 32*(5), 411–428.

Guillén, M. F. (2001). Is Globalization Civilizing, Destructive or Feeble? A Critique of Five Key Debates in the Social Science Literature. *Annual Review of Sociology, 27*, 235–260.

Hall, P. (1989). *The Political Power of Economic Ideas: Keynesianism Aross Nations*. Princeton: Princeton University Press.

Hall, P., & Taylor, R. (1996). Political Science and the Three New Institutionalisms. *Political Studies, 44*(5), 936–957.

Hall, P. A., & Soskice, D. (2011). *Varieties of Capitalism: The Institutional Foundations of Comparative Advantage*. New York: Oxford University Press.

Hamilton, G., & Biggart, N. W. (1988). Market, Culture, and Authority: A Comparative Analysis of Management and Organization in the Far East. *The American Journal of Sociology, 94*(Supplement), S52–S94.

Harwit, E. (1995). *China's Automobile Industry: Policies, Problems and Prospects*. New York: M.E. Sharpe.

Held, D., McGrew, A., Goldblatt, D., & Perraton, J. (1999). *Global Transformations: Politics, Economics and Culture*. Stanford: Stanford University Press.

Hirst, P., & Thompson, G. (1999). *Globalization in Question: The International Economy and the Possibilities of Governance*. Cambridge, MA: Blackwell Publishers.

Humphrey, J., & Schmitz, H. (2002). How Does Insertion in Global Value Chains Affect Upgrading in Industrial Clusters? *Regional Studies, 36*(9), 1017–1027.

Hymer, S. H. (1976). *The International Operations of National Firms: A Study of Direct Foreign Investment*. Cambridge, MA: MIT Press.

Iwasaki, I., & Suzuki, T. (2016). Radicalism Versus Gradualism: An Analytical Survey of the Transition Strategy Debate. *Journal of Economic Surveys, 30*(4), 807–834.

Jacobsen, J. K. (1995). Much Ado About Ideas: The Cognitive Factor in Economic Policy. *World Politics, 47*(2), 283–310.

Johnson, C. (1982). *MITI and the Japanese Miracle: The Growth of Industrial Policy, 1925–1975*. Stanford: Stanford University Press.

Kaplinsky, R. (2000). Globalisation and Unequalisation: What Can Be Learned from Value Chain Analysis? *Journal of Development Studies, 37*(2), 117–146.

Koeller, C. T. (1995). Innovation, Market Structure and Firm Size: A Simultaneous Equations Model. *Managerial and Decision Economics, 16*(3), 259–269.

Kogut, B. (1984). Normative Observations on the International Value-Added Chain and Strategic Groups. *Journal of International Business Studies, 15*(2), 151–167.

Kornai, J. (1979). Resource-Constrained Versus Demand-Constrained Systems. *Econometrica, 47*(4), 801–819.

Kornai, J. (1986). The Soft Budget Constraint. *Kyklos, 39*(1), 3–30.

Kornai, J. (1992). *The Socialist System*. Princeton: Princeton University Press.

Liew, L. H. (1995). Gradualism in China's Economic Reform and the Role for a Strong Central State. *Journal of Economic Issues, 29*(3), 883–895.

Lin, J. Y. (2012). *New Structural Economics: A Framework for Rethinking Development*. The World Bank. Washington, DC.

Link, A. N. (1980). Firm Size and Efficient Entrepreneurial Activity: A Reformulation of the Schumpeterian Hypothesis. *Journal of Political Economy, 88*, 771–782.

Mahoney, J. (2000). Path Dependence in Historical Sociology. *Theory and Society, 29*, 507–548.

Mann, J. (1997). *Beijing Jeep: A Case Study of Western Business in China*. Boulder: Westview Press.

McMichael, P. (2003). *Development and Social Change: A Global Perspective*. Thousand Oaks: Pine Forge Press.

Naughton, B. (1994a). *Growing Out of the Plan: Chinese Economic Reform, 1978–1993*. New York: Cambridge University Press.

Naughton, B. (1994b). Chinese Institutional Innovation and Privatization from Below. *The American Economic Review, 84*(2), 266–270.

Nee, V. (1989). A Theory of Market Transition: From Redistribution to Markets in State Socialism. *American Sociological Review, 54*(5), 663–681.

Nee, V. (1992). Organizational Dynamics of Market Transition: Hybrid Forms, Property Rights, and Mixed Economy in China. *Administrative Science Quarterly, 37*(1), 1–27.

Nelson, R. R., & Winter, S. G. (1982). *An Evolutionary Theory of Economic Change*. Cambridge, MA: Belknap Press.

Nicholas, T. (2003). Why Schumpeter Was Right: Innovation, Market Power, and Creative Destruction in 1920s America. *The Journal of Economic History, 63*(4), 1024–1058.

North, D. C. (1990). *Institutions, Institutional Change and Economic Performance*. Cambridge, UK: Cambridge University Press.

Oi, J. C. (1992). Fiscal Reform and the Economic Foundations of Local State Corporation in China. *World Politics, 45*(1), 99–126.

Oi, J. C. (1996). *Rural China Takes Off: Incentives for Industrialization*. Berkeley: University of California Press.

Panic, M. (2007). Does Europe Need Neoliberal Reforms? *Cambridge Journal of Economics, 31*(1), 145–169.

Panitch, L. (1995). Globalization and the State. In R. Miliband & L. Panitch (Eds.), *Socialist Register 1994: Between Globalism and Nationalism* (pp. 60–93). London: Merlin Press.

Pereira, A. A. (2008). Whither the Developmental State? Explaining Singapore's Continued Developmentalism. *Third World Quarterly, 29*(6), 1189–1203.

Perraton, J., Goldblatt, D., Held, D., & McGrew, A. (1997). The Globalization of Economic Activities. *New Political Economy, 2*(2), 257–277.

Porter, M. E. (1985). *Competitive Advantage*. New York: Free Press.

Qian, Y., & Xu, C. (1993). Why China's Economic Reforms Differ: The M-Form Hierarchy and Entry/Expansion of the Non-State Sector. *Economics of Transition, 1*(2), 135–170.

Roaf, J., Atoyan, R., Joshi, B., & Krogulski, K. (2014). *25 Years of Transition: Post-Communist Europe and the IMF*. Washington, DC: International Monetary Fund.

Robinson, W. (2001). Social Theory and Globalization: The Rise of a Transnational State. *Theory and Society, 30*, 157–200.

Ruggie, J. G. (1995). At Home Aboard, Aboard at Home: International Liberalisation and Domestic Stability in the New World Economy. *Millennium: Journal of International Studies, 24*, 507–526.

Sachs, J. (1989). My Plan for Poland. *International Economics, 3*(6), 24–29.

Sachs, J. (1993). *Poland's Jump to the Market Economy*. Cambridge, MA: MIT Press.

Sassen, S. (1996). *Losing Control? Sovereignty in an Age of Globalization*. New York: Columbia University Press.

Scherer, F. (1965). Firm Size, Market Structure, Opportunity and the Output of Patented Inventions. *American Economic Review, 55*(5), 1097–1123.

Schumpeter, J. A. (1942). *Capitalism, Socialism and Democracy*. New York: Harper.

State Council of China. (1994a). *Outlines of the Industry Policy in the 1990s*. State Council of China. (in Chinese).

State Council of China. (1994b). *The Industry Policy of the Automobile Industry*. State Council of China. (in Chinese).

State Council of China. (2004). *The Industry Policy of the Automobile Industry*. State Council of China. (in Chinese).

Steger, M. (2005). Ideology of Globalization. *Journal of Political Ideologies, 10*, 11–30.

Strange, S. (1996). *The Retreat of the State: The Diffusion of Power in World Economy*. New York: Cambridge University Press.

Sturgeon, T., Van Biesebroeck, G., & Gereffi, G. (2008). Value Chains, Networks and Clusters: Reframing the Global Automotive Industry. *Journal of Economic Geography, 8*, 297–321.

Symeonidis, G. (1996). *Innovation, Firm Size and Market Structure: Schumpeterian Hypotheses and Some New Themes*. OECD Economics Department Working Papers, No. 161.

Thelen, K. (1999). The Historical Institutionalism in Comparative Politics. *Annual Review of Political Science, 2*, 369–404.
Thun, E. (2006). *Changing Lanes in China: Foreign Direct Investment, Local Governments, and Auto Sector Development*. New York: Cambridge University Press.
Vernon, R. (1971). *Sovereignty at Bay*. New York: Basic Books.
Wade, R. (1990). *Governing the Market: Economic Theory and the Role of the Government in East Asian Industrialization*. Princeton: Princeton University Press.
Wade, R. (1996). Globalization and Its Limits: Reports of the Death of the National Economy Are Greatly Exaggerated. In B. Suzanne & R. Dore (Eds.), *National Diversity and Global Capitalism* (pp. 60–88). New York: Cornell University Press.
Walder, A. (1995). Local Government as Industrial Firms. *American Journal of Sociology, 101*(2), 263–301.
Wallerstein, I. (1974). *The Modern World-System, Vol. I: Capitalist Agriculture and the Origin of the European World-Economy in the Sixteenth Century*. New York: Academic Press.
Wallerstein, I. (1980). *The Modern World-System, Vol. II: Mercantilism and the Consolidation of the European World-Economy, 1600–1750*. New York: Academic Press.
Wallerstein, I. (1989). *The Modern World-System, Vol. III: The Second Era of Great Expansion of the Capitalist World-Economy, 1730–1840s*. New York: Academic Press.
Walter, A. (2006). From Development to Regulatory State? Japan's New Financial Regulatory System. *The Pacific Review, 19*(4), 405–428.
Weber, M. (1978). *Economy and Society: An Outline of Interpretive Sociology*. Berkley: University of California Press.
Weiss, L. (1998). *Myth of the Powerless State*. Ithaca: Cornell University Press.
Williamson, O. E. (1993). Transaction Cost Economics and Organization Theory. *Industrial and Corporate Change, 2*(2), 107–156.
Winter, S. G. (1984). Schumpeterian Competition in Alternative Technological Regimes. *Journal of Economic Behavior and Organization, 5*, 287–320.
Woo, W. T. (1999). The Real Reasons for China's Growth. *The China Journal, 41*, 115–137.
Yeung, H. W. (2000). State Intervention and Neoliberalism in the Globalizing World Economy: Lessons from Singapore's Regionalization Programme. *The Pacific Review, 13*(1), 133–162.
Yeung, H. W., & Coe, N. M. (2015). Toward a Dynamic Theory of Global Production Networks. *Economic Geography, 91*(1), 29–58.

CHAPTER 2

FAW, "Senior Son" of the Chinese Automobile Industry

Located in Changchun city, Jilin Province, the heart of northeast China, First Auto Works (FAW) was the first automobile factory established by the central government of the People's Republic of China. Owing to the national importance consistently accorded to it by the state, this auto-giant is commonly portrayed as the "senior son" of the Chinese automobile industry. A flagship project of China's former planned economy, FAW had from the outset belonged to the central government and strictly followed economic commands from above. Despite the profound changes to the rigid state–enterprise relationships that followed the market reform, the central government maintained a firm grasp on the prized FAW, viewing it as a key element in the state's control apparatus over the economy, while ideologically entreating FAW to fulfill the national will. The organization of production at FAW has thus, from its establishment in 1950, and throughout the far-reaching economic changes in China, remained deeply connected to the leading role of the central government. This chapter explores FAW as an exampe of the first production model, showing the institutional reform of a state-owned and managed enterprise, its subsequent adaptation to changing market conditions and its ultimate engagement with foreign partners to foster the advances required for the mass production of modern cars.

The following sections first review the pre-reform economic positioning of the northeast region as the national heavy-industry base, with particular focus on the organizational features of FAW under the centralized

Q. Feng, *Variety of Development*,
https://doi.org/10.1007/978-981-10-5912-4_2

communist regime. The chapter then focuses on the changing local political structure and configuration of developmental ideas arising from China's market transition, during which FAW's legacy as a state-directed automobile enterprise was ensured by the central government's decision to retain certain key enterprises under its direct control, and its subsequent pioneering role in developing new models of enterprise governance. In these ways, as the new market environment began to pose challenges to the Chinese automobile industry during the 1980s, central policymakers mobilized FAW as an instrument for trialing emerging developmental ideas. Accordingly, this chapter delves into the socio-political process of the internal transformation of FAW amid the wider upheavals of market reforms, which saw the state-owned factory evolve into a corporation directly owned by the central government as the head of a large corporate group, which then sought out joint ventures with foreign partners as a strategy for technological upgrading.

Northeast China: The National Heavy-Industry Base

Ever since communist leaders gained political control and began to imagine a new China in the 1950s, the expansion of heavy industry had been a national economic priority. With substantial existing infrastructure compared to other regions, and close access to the border with its then Soviet allies, China's northeast provinces of Harbin, Jilin and Liaoning were chosen as the location that would pave the way for the country's industrial future. The ensuing renewal of the northeast as the national heavy-industry base during the communist period resulted from a colossal planning and construction led by the central government. Through the combination of intensive national investment in local industrial projects, establishment of institutions and regulatory regimes to govern local economic activities, and roll-out of numerous local enterprises under the control of central government agencies, the northeast heavy-industry base epitomizes the functioning of China's planned economy.

The Chosen Land

The ardent pursuit of heavy industry was deeply rooted in the political philosophy of the new China, exemplified by Chairman Mao's famous complaint about the underdevelopment of Chinese heavy industry: "What can we make now? Tables, chairs, tea bowls, tea pots, foodstuffs, flour powder and paper, but not a car, an aircraft, a tank or a tractor." Accordingly, when national planners drafted China's first five-year plan (1953–1957), an explicit guideline was included to prioritize investment in heavy industry over all other sectors, underscoring projects related to steel, coal, electricity, petroleum, machinery, non-ferrous metals and chemicals. The primary content of the first five-year plan comprised "156 projects" aided by the Soviet Union, 147 of which were in heavy industry,[1] and intensive industrial investment continued throughout the entire period of the planned economy. From 1953 to 1978, the annual growth of the heavy sector was maintained at an average rate of 13.8%, significantly higher than 11.4% for industry overall and 2.7% for agriculture (National Statistics Bureau 1999). In order to mobilize national resources such as financial capital, raw materials and labor for the heavy sector, the state designed a series of institutional mechanisms which formed the skeleton of China's planned economy and which arguably had a more profound impact than investments alone. A hierarchical bureaucratic system, consisting of a National Planning Committee and its sub-divisions, was established to oversee economic plans and resource allocation; a state-owned banking system was developed to raise funds for industrial projects; and critical materials such as coal, steel, cement, and wood were tightly controlled by the National Bureau of Materials to enforce national economic plans. Meanwhile agricultural products were purchased and sold by the government at low prices to subsidize industrial development, and a national system of human resource management was developed, in which students, workers, veterans, and cadres were educated, trained and assigned jobs based on national planning.

In keeping with these ideas, during the three decades of China's planned economy, heavy industry expanded across all regions of the country; however, for central planners, the northeast was always the priority location for key projects. As early as 1945, at the 7th National Congress of the Communist Party of China (CPC), Chairman Mao urged his followers to quickly occupy the northeast, since this region of China had excellent industrial infrastructure and easy access to the Soviet Union, thus offering a solid base for the future communist revolution. In the ensuing civil war (1945–1949), the CPC concentrated most of its resources on establishing control over this region, as well as taking advantage of local industries to supply its military hardware. In March 1950, a year after the establishment of the new China, Chairman Mao held a conference in Shenyang, a major city of the northeast, at which he re-emphasized the role of the northeast as the national heavy-industry base, and his expectation that this region would provide the "machines" and "experts" needed for the whole nation (Fu 2004).

What made China's northeast so special in the eyes of national leaders and planners? One major reason was its existing industrial foundation, historically established by the Japanese colonizing force, for whom the northeast was intended to be a permanent industrial base serving their grand invasion strategy of China. From the 1930s, as Japan gradually expanded control over the northeast via the puppet Manchurian government, until Japan's surrender at the end of World War II, huge sums of Japanese capital were invested in the region for heavy industrial projects. From 1932 to 1936, Japanese investment totaled 1.15 billion yen, 98.9% of all foreign investments in this area. From 1937 to 1941, Japanese investment in the region increased to 4.7 billion yen, finally reaching over 10.7 billion yen by 1945 (Zheng 2000). These massive investments accelerated the industrialization of the northeast. From 1932 to 1944, local coal output more than trebled from 7.09 million to 25.62 million tons, steel output quadrupled from 0.22 million to 0.88 million tons, and there was similar growth in the petroleum, chemicals, electricity, and non-ferrous metal sectors (Yi and Lin 2001). As a consequence, the northeast quickly became the most advanced site of heavy industry in China. By 1943, the region accounted for approximately 49% of the total national output of coal, 60% of barilla, 69% of vitriol, 88% of raw iron, 93% of steel, 93% of electricity, and 95% of machinery (Yi and Lin 2001).

After the victory of the Chinese communists in 1949, the already strategically valuable northeast was thus earmarked as the national base of heavy industry, and the regional economy started to be shaped by the templates of central planners. State-led industrial expansion began immediately. In 1950, about half of the total national investment and half of China's college graduates were directed to the northeast; and in 1951, about two-fifths of national infrastructure construction took place in this region (Fu 2004). During the first five-year plan, the central government assigned 57 of "156 projects" aided by the Soviet Union to the northeast. Among the non-military projects, the three northeast provinces accounted for over 50 projects, approximately half of the total national investment (Table 2.1). The concentration of national investment and assignment of key industrial projects to this region continued over later periods, and until China's market reforms, the northeast remained the focal area for investments from the central authorities.

With heavy industry comprehensively established in the northeast, the central government fostered a cluster of national top industrial enterprises in the region, including Daqing Oilfield, Anshan Steel and Iron Works, Shenyang Aircraft, and FAW. Owned and directed by the central government,

Table 2.1 Distribution of non-military projects in "156 projects" across provinces of China

	Total projects	*Planned investment*		*Actual investment*	
		amount (1 million yuan)	*National proportion*	*amount (1 million yuan)*	*National proportion*
Northeast provinces					
Liaoning	20	4.07	25.2%	4.59	28.9%
Heilongjiang	20	1.77	11.0%	2.05	12.9%
Jilin	10	1.36	8.5%	1.46	9.2%
Subtotal	50	7.20	44.7%	8.10	51%
Other provinces					
Henan	9	2.54	15.7	1.55	9.8
Gansu	7	1.30	8.1	1.31	8.3
Shanxi	7	0.54	3.3	0.57	3.6
Shaanxi	7	0.40	2.5	0.43	2.7
Hebei	5	0.28	1.7	0.28	1.8
Subtotal	35	5.06	31.3%	4.14	26.2%

Source: Based on Dong (2004)

Note: Only provinces assigned over five projects are listed

these enterprises provided significant returns to the nation. In the first five-year period, the financial return from Heilongjiang province to the central government was over three times the original sum invested in local industry; for Jilin, the fiscal turnover during the same period was 1.66 billion yuan, about 60% of the total provincial income; while industrial enterprises in Liaoning created a return of about four times the original investment from 1953 to 1988 (Yang 2005).

Arguably of equal importance, industrial expansion in the northeast allowed the state apparatus of the new China to trial mechanisms for governing enterprises in a planned economy. Unlike corporations in a market economy, Chinese state-owned enterprises operated more like order-receiving factories, with no autonomy outside economic commands received from above. The state usually built infrastructure, introduced technology and equipment, trained the labor force, appointed directors, issued production plans, supplied raw materials, and took over output and profits, leaving nothing for the factories themselves except to follow instructions and accomplish production assignments. Within these factories, a "director-responsibility system under party-committee direction" was developed (see Fig. 2.1). For a typical factory, the top bureaucratic

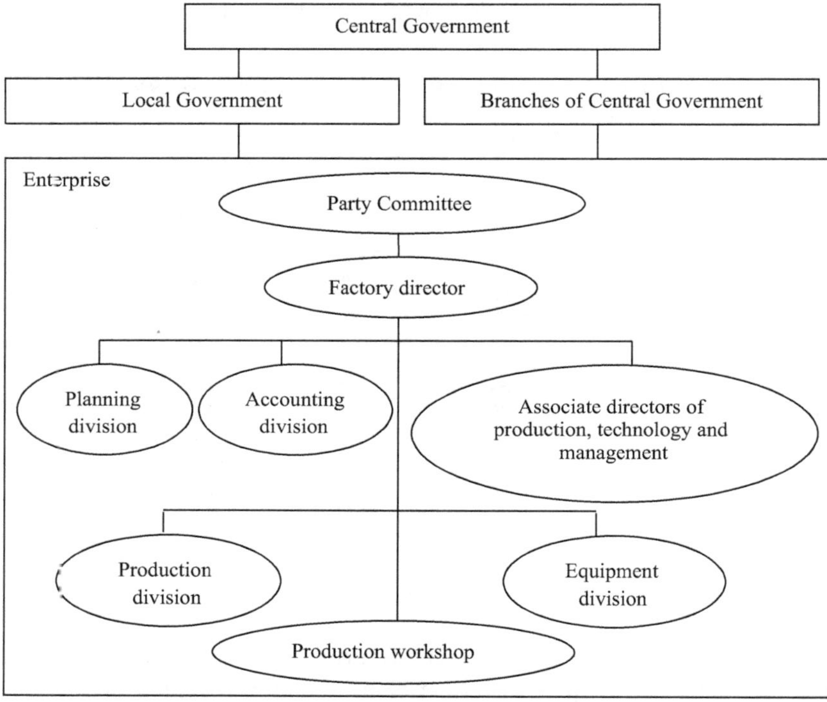

Fig. 2.1 Governance of state-owned enterprise in the planned economy of China
Source: Based on Ma and Liu 2000

unit was the party committee, beneath which the factory director was in charge of ordinary management. Production was usually carried out in multiple workshops. These external and internal mechanisms of enterprise governance required that state-owned enterprises were loyal executors of hierarchically assigned economic plans, operating like fingers on the hand of the central government.

This mode of governance established the enterprises of the northeast as national role models. In its early stages, the Chinese planned economy had an acute shortage of actual enterprise management experience, and especially of qualified managers; political elites within the party committees were often appointed to manage enterprises. The northeast soon became the de facto "management school", training a successive cohort of managers that gradually populated senior positions across

China. Moreover, major innovations in management were often exported by the northeast enterprises to other regions and sectors. For instance, in 1960, Anshan Steel and Iron Works in the northeast replaced the previous manager-centered scheme with a new system featuring mass participation of workers in management and a close bond between managers, technicians, and workers in production, an innovation that later generated nationwide impacts. In the mid-1960s, Daqing Oilfield, a major northeast enterprise, was held up by the central government as another laudable management model, featuring a combination of communist enthusiasm for scientific management approaches, the popularization of model workers in production, and insistence on discipline and responsibility.

FAW: "Best Specimen" of the Planned Economy

As the first and largest automobile enterprise in the new China, FAW was a model for operation of a state-owned enterprise under central government direction. For Gen Shaojie, FAW's director in the 1980s and 1990s, this enterprise was a "specimen" and a "perfect product" of the planned economy (Wang 1998), an apt description as regards the pre-reform history of FAW. In 1950, just a year after the establishment of the new China, the Ministry of Heavy Industry began the search for locations for its first automobile factory. In 1951, Zhou Enlai, prime minister of the State Council, proposed the city of Changchun in Jilin Province, and this was soon confirmed by the Ministry of Heavy Industry and the State Council. In 1953, in a special announcement on FAW, Chairman Mao declared that the construction of related infrastructure should be completed within three years. This was the first time the central government had made such a pronouncement about an individual enterprise, and it included the following instruction to the nation (First Auto Works 2003a: 8):

> As our technologies are underdeveloped and experiences are limited, to accomplish such a large project in only three years, we will encounter major difficulties in labor force recruitment, construction technologies, supply of domestic equipment, and production preparation. Therefore, the central government considered it necessary to make an announcement to the whole country, and to request all related institutions to provide their maximum support to this automobile factory in Changchun. Try the best to accomplish it within three years.

With this order from the central leadership, construction of FAW became a nationwide priority and resources from all over China were soon mobilized towards the city of Changchun in Jilin Province. Machinery, equipment, and construction materials from about 70 domestic enterprises were transported to Changchun; and all Chinese provinces except Tibet and Qinghai sent workers, technicians, experts, and managers to assist the project (First Auto Works 2003b). The FAW facility was successfully completed on schedule in 1956, with local government agencies acting as key assistants under the direction of central planners. Both the Jilin provincial government and the municipal government of Changchun were required by the national leaders to guarantee the success of the FAW project. In Changchun, 95% of the municipal construction budget in 1954 and 85% in 1955 were set aside for FAW infrastructure, including roads, bridges, and tunnels for water and gas (First Auto Works 2003b). In 1956, the party committee of Jilin Province issued a robust statement on the duties of the local governments in relation to FAW:

> (1) The task of this automobile factory is to comprehensively fulfill the commands of the central committee, to warrant the accomplishment of planned assignments and commands from the First Machinery Department, and also to assure manufacture of quality products of domestic-made automobiles and cultivation of qualified personnel for the Chinese automobile industry; (2) The municipal party committee of Changchun should strengthen their supervision on the party committee of FAW, and organize local resources to fully assist FAW, warranting the accomplishment of production tasks through political and ideological works; (3) Related institutions of the whole province should keep strengthening their assistance on the FAW project. Factories coordinating with the production of FAW and their local party committees at the municipal, county and district levels should guarantee production standards, product quality, and the timeliness of fulfilling production contracts. Related institutions in charge of railways, business, labor force, transportation, supply of materials, culture, and public health should all perform well in assisting FAW. (First Auto Works 2003a: 23)

The technologies required by FAW were mainly supplied by the Soviet Union under central government arrangements set out in "156 projects". Stalin Automobile Factory, then the largest automotive factory in the Soviet Union, played a key role, setting up a specific division in charge of working with the FAW project, developing technological and design aspects, providing production equipment and tools, and coordinating with other assisting factories in the Soviet Union. With the approval of Stalin Automobile

Factory, FAW also acquired an important handbook, *Design of Production Organization*, a key reference for managing the automobile enterprise that was soon adopted nationwide in China as the major management textbook in related sectors. Sino-Soviet technology transfers were accomplished through the exchange of technicians. From FAW, eight technician teams totaling 518 individuals were sent for internships with Stalin Automobile Factory, in which they were carefully assigned to various positions in production and management; in the other direction, from 1953 to 1957 the Soviet Union sent 188 experts to FAW, providing over 20,000 recommendations and giving 1300 lectures (First Auto Works, 2003b). In consequence, Stalin Automobile Factory was widely considered at FAW as the "mother factory", which "produced an automobile industry for China" (Sino-Soviet Friendship Association in Jilin and Changchun 1956: 15).

Equipped with the necessary infrastructure, technology, and trained human resources, FAW started to operate as an exemplary state-owned enterprise in the planned economy, developing and practicing the so-called director-responsibility system under party-committee direction. As mentioned, under this model, party institutions directly oversaw FAW management divisions, which in turn managed production branches according to the commands of the central government (see Fig. 2.2). It was a scheme of governance that axiomatically placed most management activities under the direction and monitoring of the government, leaving minimal autonomy for managers. Liu Shouhua, factory director of FAW for 16 years during the planned-economy period, once commented that, as almost everything at FAW had been planned and approved from above, he only had control over minor decisions such as construction projects of less than 100 square meters. As a result, a popular joke was that he only had the power to build a new toilet block. From 1953 to 1983, FAW fulfilled its assigned duties with an output of 1.02 million vehicles and a significant fiscal turnover of about 5.1 billion yuan, approximately eight times the original investment (Xu 1983). Regarding product development, to serve national industrialization plans, the state designated medium-sized trucks as the priority product at FAW. A 4-ton truck model introduced from Soviet vehicle designs, branded the CA10 Jiefang ("liberation" in Chinese), became the mainstay of FAW production until the start of the market reform.

Only a small quantity of passenger cars was produced at FAW for formal state use in this period of time. The initial idea of producing passenger cars can be traced back to Chairman Mao, who expressed an expectation in 1956 to "sit in our own cars". The CA71, or Dongfeng ("east wind" in Chinese, a symbol of communism) was China's first car model, quickly

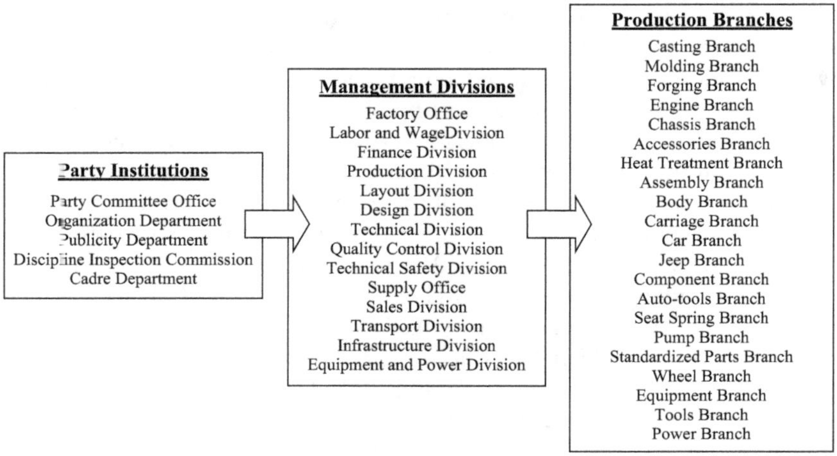

Fig. 2.2 Organizational structure of FAW under the planned economy in 1979
Source: Based on Historical Records Office of First Auto Works 1991

produced by FAW for delivery to Chairman Mao for a trial ride in 1958. A larger model CA72, or Hongqi ("red flag", symbolizing revolutionary spirit) was later developed in 1959 as a political gift for the ten-year anniversary of the new China. The Hongqi CA770, an upgraded version of the CA72, was developed in 1965 and widely utilized in major political events as "the nation's car". In fact, these cars extensively imitated foreign models: the Dongfeng used the Mercedes-Benz engine and largely replicated the French Simca; the Hongqi CA72 was modelled on a Chrysler; while the Hongqi CA770 design referenced Mercedes-Benz and Rolls-Royce models. However, being mainly aimed at political use, these cars were never mass produced. The total outputs of Hongqi CA72 and CA770 models were only 202 units and 1520 units, respectively, before production was suspended in 1981 (Historical Records Office of First Auto Works 1998).

State–Enterprise Relationship in Reform: Dual-Track Transitions

China's market reform, set in motion from the early 1980s, altered the planned-economy system and prompted comprehensive changes in the country's economic institutions. The state–enterprise relationship in the northeast, in which the dominant role had been played by central authorities, underwent a dual-track transition. On the one hand, market reforms gradually released most enterprises from the grip of direct government control, and former state-run factories evolved into autonomous corporations in the newly created market economy. In parallel, the central government decided to retain control of a few carefully selected enterprises, to maintain its influence over key sectors, and by extension, the broader national economy. FAW, having been heavily favored at the outset of the planned economy due to the northeast's economic positioning, was among those chosen by the government to retain its control during the reforms. Through this dual-track market transition, the dominance of the central government over selected enterprises such as FAW, far from diminishing, actually endured, albeit taking on different forms.

The major landmark that set China on the path to reform was the 1978 Third Plenary Session of the 11th Central Committee of the CPC. Two subsequent waves of institutional reform sought to reshape the relationship between government and enterprise in China. These reforms were based on a strong acknowledgement among China's central leaders that the over-centralized governance structure had become a serious problem for the national economy, and it was thus necessary to transfer more autonomy to enterprises to foster and adapt to a new market environment.

The first wave of reform, from 1979 to 1992, was characterized by the marked decline of bureaucratic intervention in ordinary enterprise management. In the first half of the 1980s, a series of administrative rights, previously held by governmental planning commissions, had been reassigned downwards to enterprise level. Enterprises began to enjoy a free hand in decisions relating to procurement, employment, production, sales, and pricing, and most importantly, a proportion of profits were allowed to be retained. Starting from 1986, a responsibility system was introduced as a further encouragement for enterprises to be more independent from government control. In a typical responsibility system, instead of receiving direct bureaucratic orders, enterprises were asked to sign production contracts with the state which required them to be responsible for any loss, while allowing them to retain a certain rate of profits. In 1988, the central government issued the Law on State-Owned Industrial Enterprises, which listed 13 major rights for state-owned enterprises and formally confirmed the responsibility system. Meanwhile, the State Council issued a set of guidelines on the internal management of state-owned industrial enterprises, which permanently replaced the old director-responsibility system under party-committee direction. Under the new management framework, the administrative obligations and rights of enterprises were more clearly delimited, and the leading role of the enterprise director in relation to ordinary management procedures was confirmed.

The increasing autonomy of state-owned enterprises encountered another breakthrough in 1992 with the launch of a second wave of reforms aimed at establishing "modern enterprise institutions" by "corporatizing" state-owned enterprises. Four principles were set out by central government in 1993 to guide the new wave of institutional reforms and remold the state–enterprise relationship: clarification of the property rights of enterprises; clarification of the overall rights and responsibilities of enterprises; separation of the state and enterprise; and promotion of "scientific" management in enterprises. To address the previously ambiguous property rights of state-owned enterprises was, however, the central mission of this process. The 1993 Third Plenary Session of the 14th Central Committee of the CPC unprecedentedly acknowledged the corporate system and the shareholding system. In the same year, China's first company law was issued. In 1997, the shareholding system was further legitimized, encouraging the coexistence of multiple forms of ownership while simultaneously ensuring the dominance of public ownership in China. These institutional changes created a major "corporatization" and "equitization" movement among Chinese state-owned enterprises.

Despite steering the state–enterprise relationship towards a market model, the ongoing concern for reformers was to maintain the necessary influence of the central government over the national economy in the new market environment. In the 1980s and early 1990s, as state-owned enterprises were gradually released into the market, for political reasons, central leaders always insisted on retaining ownership. To maintain the predominant position of public ownership in China was considered a symbolic indicator of both the political validity of the reforms and the legitimacy of the emerging market economy. However, government insistence on maintaining ownership faced increasing troubles as marketization accelerated over the 1990s. The large number of state-owned enterprises started to impose a major burden on the Chinese economy: they dispersed the financial resources of the state, generated redundant industrial projects, reduced production scale and efficiency, delayed technological upgrading, and accumulated heavy losses and debts. Facing these intractable problems, some policymakers complained that the government "was not capable to manage, and was also not capable to manage well" so many state-owned enterprises (Li 1996: 6).

In response, from the mid-1990s the state began to experiment with new governance approaches that aimed to reduce the share of public ownership of enterprises while retaining the state's influence over the market. Some state-owned enterprises were abandoned and the government concentrated on a few large and significant ones: the "strategic transformation of the state-owned enterprises". Under this new policy orientation, non-state firms were encouraged to take over small state-owned enterprises, transforming many enterprises from state ownership to a non-state category. At the Fourth Plenary Session of the 15th Central Committee of the CPC in 1999, the central leaders established more systematic principles with regard to the role of the government in the national economy, set out as follows:

> Under the socialist market economy, the leading role of the state-owned economy in the national economy should be mainly reflected by its controlling power. (1) The role of the state-owned economy should be enforced not only through solely state-owned enterprises, but also through state-holding enterprises and enterprises with state-owned shares based on development of the shareholding system; (2) The state-owned economy should hold a predominant position in focal sectors and key fields of the national economy, should support, lead, and drive overall socioeconomic development, and should play a vital role in realizing the goal of national macroeconomic-control; (3) The state-owned economy should maintain at a necessary quantity, and more importantly, should develop a better layout and quality. (Central Committee of the Communist Party of China 1999)

As guided by the second principle above, the state decided to concentrate its control over several key sectors, considered the chief apparatus through which the broader national economy would be controlled. These key sectors were those relating to national security, natural-monopoly industries, critical public goods and services, high-tech industries, and the so-called pillar industries. Of the latter, the automobile industry had long been deemed critical to the national economy by central policymakers, undoubtedly not one to be released. Accordingly, FAW as the earliest and largest enterprise in the Chinese automobile sector was designated a "significant" enterprise to be retained under the control of central planners.

However, profound changes were under way in specific governance mechanisms. The old government bureaucratic institutions in charge of state-owned enterprises were being either dismantled or transformed. The largest round of such changes occurred in 1998, when forty departments of the State Council were reduced to twenty-nine, and about 200 lower-level institutions were abolished entirely. In particular, nine sector-specific bureaus were incorporated into the National Economic and Trade

Commission, and thus permanently lost their privileges over related enterprises. These reforms ended the direct interventions of the central administration into the operations of enterprises, making state interventions ever more reliant on industrial policy directives. By the early 1990s, the central government had issued numerous industrial policies with comprehensive coverage of all economic sectors, to which the National Outline of Industrial Policy in the 1990s, issued in 1994, served as a guide, detailing how the government should now direct the national economy (see State Council of China 1994a). This document directed that industrial policy should be implemented mainly through economic measures, legal measures and "necessary" administrative measures, rather than bureaucratic orders as practiced before. Though the deliberate opaqueness of the "necessary" administrative measures still suggested an interventionist tendency, this document represented major progress in the reform of state economic governance. The same year saw the issue of the first automobile industry policy, also the first sector-specific industrial policy in China (see State Council of China 1994b).

With the increase in industrial policies, a new mechanism was developed in 1999 to maintain the state's control over important enterprises such as FAW a specific committee in charge of about two hundred state-owned enterprises, all of which were deemed crucial to the national economy and thus not to be released to the market. In 2003, this committee became the State-Owned Asset Supervision and Administration Commission (SASAC). According to the central government, SASAC is supposed to undertake the responsibility of investors on behalf of the state, supervise and manage state-owned assets of enterprises in accordance with the relevant laws, and promote the reform and restructuring of state-owned enterprises. One major way for SASAC to enforce its will in these enterprises is to nominate and supervise their senior managers, while for the most important enterprises such as FAW, even higher approval is required from the Politburo of the Central Committee of the CPC. Currently, the SASAC retains about one hundred such "central enterprises".

New National Orientation for the Car Industry

The automobile manufacturing sector further gained the special attention of the central government following the market reform, owing in part to the radically shifting domestic economy. Rapid economic growth in China after the 1980s led to an unexpected explosion of market demand for automobiles, especially passenger cars, and the corresponding influx of foreign vehicles soon became a major headache for central policymakers. Import substitution was thus perceived as an urgent mission for Chinese automobile enterprises. As policymakers battled to find solutions in the face of wave after wave of imported cars, setting up joint ventures with foreign corporations was finally proposed as a feasible way to upgrade the technology of Chinese automakers and expand domestic production as quickly as possible. These new ideas on managing the auto sector heralded a new orientation for Chinese automobile manufacturing, and laid out a specific development pathway for large domestic automakers, with FAW once again at the forefront.

In the early days of the market reform, the importance of the automobile sector was much debated among policymakers. The shattered national economy that resulted from the Cultural Revolution (1966–1976), coupled with the consequent loss of political legitimacy for the CPC, impelled political leaders to set economic development as the top priority in the reform. Yet, the automobile sector was not originally perceived as important to national economic recovery. An influential view at this time was that so long as human resources remained abundant, China might not need motorized vehicles as tools of production, and national economic planning often restricted automobile output to save resources for other more important projects. Rapid economic growth in China after the 1980s, however, led to an unexpected explosion of market demand for automobiles, especially passenger cars, and the corresponding influx of foreign vehicles soon became a major headache for central policymakers. Officials in charge of the automobile industry argued that although a structurally complete and independent domestic automobile industry had been successfully established during the planned economy, China's automotive production remained highly underdeveloped. Industry actors warned that poor transportation capacity as a result of overlooking the need to expand automobile production was becoming a bottleneck for the national economy (e.g. Li 1983).

A further critical issue was that Chinese automobile manufacturing had been severely biased with regard to product structure, being based on the limited diversity of models discussed above. The long-term focus on medium-sized trucks as the single main product, exemplified by the CA10 Jiefang which had been produced at FAW for thirty years without any major technical upgrades, had created a particular predicament, commonly described as "lacking heavies (heavy trucks), few lights (light trucks), and almost no cars". It was broadly acknowledged that Chinese automobile manufacturing technology lagged far behind international standards before the reform. Like the Soviet automobile factories and production models they copied, FAW and many other Chinese assemblers tended to place output above quality and innovation, as common adage among these factories had it: "to have it or not is the first matter, to be good or bad is secondary".

As industry supporters strenuously lobbied the central government over the value of a strong automobile sector for China, the mindset of national policymakers began to change. Two goals officially announced by the central government were first, that the industry was expected to modernize technologically, so that by the 1990s, China should be able to "produce a generation of automobile products equivalent to the international level of the 1980s" (Rao 1984: xii). Second, as stated in the seventh five-year plan (1986–1990) and reconfirmed at the Fourteenth National Congress of the CPC in 1992, the automobile sector was expected to expand its role as a "pillar industry" of the national economy, serving and facilitating other associated sectors. This designation ensured the automobile industry a priority position in the broader structure of the national economy, guaranteeing special consideration and investment from the central government.

By comparison with policy discussions on the general organization of automobile production, debates on whether China should specifically produce cars were more complex. Pre-reform output of cars had been negligible, and as a result, China had one of the lowest car ownership rates per capita in the world in the 1990s. As can be seen in Table 2.2, in 1995, for every car in China there were about five hundred people, far below the rate of about 2–10 people per car in advanced societies. In response to such a massive imbalance in the number of vehicles serving the population, some Chinese scholars considered the growth of passenger cars to be a wrong turn for China, and instead advocated the development of mass public transportation systems. For these commentators, creating a "car civilization" could waste energy, pollute the environment, bring traffic jams, and even encourage consumerism (Zheng 1994), all of which were indeed fairly accurate predictions. A large public media debate ensued throughout the mid-1990s, though no consensus was ever reached (e.g. Zheng 1996).

As discussions continued among scholars, central government's policy considerations in relation to passenger cars were increasingly driven by external pressures. As the demand for cars rapidly increased, China was flooded by imported cars, draining limited national foreign currency

Table 2.2 Country rank by people per car in 1995

Rank	*Country*	*People per car*	*Rank*	*Country*	*People per car*
1	Vietnam	950.2	21	Portugal	4.4
2	China	487.9	22	Ireland	3.8
3	India	244.9	23	Czech Republic	3.5
4	Pakistan	154.0	24	Slovak Republic	3.5
5	Philippines	118.2	25	Japan	2.9
6	Indonesia	107.9	26	Spain	2.8
7	Thailand	54.0	27	Netherlands	2.6
8	Columbia	36.5	28	Belgium	2.4
9	Turkey	21.2	29	Puerto Rico	2.4
10	Russia	15.9	30	Sweden	2.4
11	Brazil	13.2	31	France	2.3
12	Venezuela	12.6	32	United Kingdom	2.3
13	Mexico	11.2	33	New Zealand	2.2
14	Chile	10.8	34	Australia	2.1
15	Singapore	8.9	35	Austria	2.1
16	Korea	8.4	36	Canada	2.0
17	Argentina	6.7	37	Germany	2.0
18	Poland	5.5	38	Italy	1.9
19	Taiwan	5.3	39	Luxembourg	1.7
20	Hungary	4.9	40	United States	1.7

Source: Sturgeon and Florida (2000)

reserves. The speed of the influx is illustrated in Fig. 2.3, showing how China's minimal car imports of about one thousand units in 1981 had soared 100-fold to over 100,000 units by 1985. The import surge of the 1980s was suppressed by a series of government regulations, but as long as domestic production was weak, substantial imports of foreign cars would continue to haunt China. In the 1990s, when the Chinese economy entered another boom, an even more significant explosion of car imports followed (See Fig. 2.3). This time, for the CPC leadership, the influx of foreign cars was not only an economic threat to domestic industry, but a political nightmare that was hurting national dignity (Li 1998). The situation was especially intolerable for China's senior automobile bureaucrats, for whom building and maintaining a self-sufficient automobile industry had been a career-defining duty and the major achievement of the planned economy. Incredulity over surging car imports was best expressed by Rao Bin, the top administrator of the Chinese automobile industry in the 1980s, who complained that China's

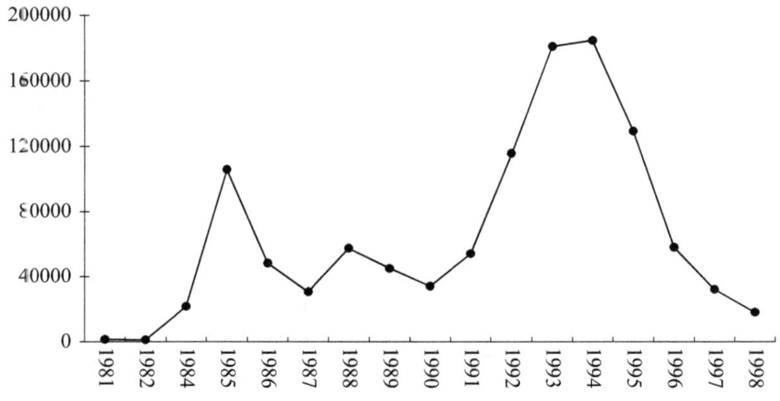

Fig. 2.3 Car imports in China, 1981–1998
Source: China Automobile Industry Yearbook 1999

expenditure on automobile imports from the 1950s had been twelve times more than the cost of building FAW (Rao 1983).

To check the perceived unrestricted car imports, the promotion of domestic production as import substitution became a key policy tenet in the 1980s and 1990s. The central government sought to suppress inflows by adopting high tariffs as the first line of defense. In consequence, tariffs on imported passenger cars were the highest of all commodities in this period. However, it was well recognized among policymakers that import regulations could not be a permanent solution, and it was imperative to promote domestic car production. In August 1984, the central government organized a discussion at Beidaihe, a summer resort, which became a milestone meeting for the re-start of the Chinese car industry. At the meeting, central leaders finally determined that domestic car production should be encouraged to meet growing domestic demand. The following year, automobile industry figures persuaded Hu Yaobang, one of the top political leaders at that time, to highlight the development of the car industry in the seventh five-year plan (1986–1990). In 1987, the state held a national conference for Chinese automakers at the Second Auto Works (SAW), later renamed Dongfeng Motor Corporation (DMC), at which central administrators began to select domestic enterprises that would become the major carriers of the car production initiative. Later the same year, the central leaders announced that domestic car production would be concentrated within three major enterprises: FAW, SAW and Shanghai Tractor and Automobile Industry Coordination Co. (STAICC, later renamed the Shanghai Automotive Industry (Group) Corporation, or SAIC), or the so-called “Big Three”. The “Small Three”—three lesser car enterprises located in Tianjin, Beijing, and Guangzhou—were added in the following year. This structure for Chinese automobile production remained unchanged until the late 1990s.

The shift in production practices needed to move towards import substitution was accompanied by a seemingly intractable associated issue: how could these government-designated enterprises possibly obtain the massive quantity of new technology required to produce cars to international production standards? With great political courage, the central government decided to overturn decades of entrenched political norms, allowing joint ventures with foreign corporations in the automobile sector. Influenced by the nationalist ideology of the Mao era, foreign capital and technology were once forbidden territory for industrial bureaucrats. Deng Xiaoping, the senior leader of the CPC in the post-Mao era and key architect of the economic reform, lifted this ideological barrier in 1978, approving the first trial car-assembly joint venture in Shanghai with Volkswagen as the foreign partner. The trial was a major success, providing solid justification for policymakers to accept full-fledged joint ventures with foreign partners to help upgrade the automobile sector. FAW’s first joint venture, FAW-VW, was started in this context. The key components required to bring the joint ventures into line with the import substitution strategy and reinforce the legitimacy of the joint-venture strategy were as follows. Central government approval for joint-venture projects insisted on a high proportion of domestic content in the manufacturing process. To retain effective control over joint ventures, foreign partners were not allowed to hold a share greater than 50%. Lastly, driven by the strong ideological legacy for a self-reliant automobile sector, policymakers envisaged that domestic automakers should eventually become technologically independent from their foreign partners. The requirements of domestic content, shared control over the joint venture projects, and the anticipated technological independence of domestic facilities reinforced the legitimacy of the joint venture strategy in the mindset of the policymakers.

These policy considerations formed the three key aspects of the Chinese automobile industry policy issued in 1994 (see State Council of China 1994b): to facilitate import substitution through high tariffs; to upgrade obsolete technologies through joint venture with foreign companies; and to concentrate the fragmented sector through entry regulations. The official belief was that core operations should be assigned to a few significant enterprises, which saw selection as the main players on the national team as an opportunity for post-reform development. The chosen few, including FAW, soon matured as privileged enterprises in the domestic market, with an official duty as practitioners of national industrial policy.

FAW: Developed Under the National Will

The central theme of all significant changes within FAW after the market reform was to serve the national will in relation to the new direction of the automobile industry. Driven by changing developmental ideas, the central government launched rounds of trials to re-build the old state factory and create a state-owned corporation that established joint ventures with foreign corporations. Similar to the early years when FAW was built up as the national factor of automobiles, the ownership structure and technological strategy of FAW in the new market environment was once again forged under central authority.

From State Factory to State-Owned Corporation

FAW was in the front line of the state's new plans to transform economic organizations. One of the central government's key ideas, to develop large enterprises and release small ones, was fully applied to the case of FAW, which was considered too significant to let go. Unlike in many other transition economies, privatization was never a preferred option for China's central leaders. During the reform, FAW, like many other old state factories, received greater autonomy from above, and adaptation to the emerging market led the factory to gradually mature as a corporation in the modern sense. Nevertheless, it maintained its position as one of the central enterprises throughout the reform period, so that central government never fully released its overarching control, albeit exercising it via a new governance system.

Reflecting the two stages of enterprise reform described above, the organizational transformation of FAW can also be divided into two periods. The first covered the process by which FAW began to gain more autonomy from the previous mode of centralized management, beginning as early as 1979. By that point, FAW had already been selected by the central government to take part in a trial of expanded rights over financial resources (FAW 2003a).

Prior to the reform, the only funds FAW could reserve for its own use came from the employee salary pool, that is, the reallocation of workers' earnings according to the economic needs of the enterprise. During the reform, however, FAW was allowed to begin extracting profits directly from production after contributing an assigned rate to the government. More specifically, the central government asked FAW to classify its overall profits into two parts, baseline and extra profit, and then set specific ratios for these two types of profit to calculate the amount due to the government. Under this new scheme, FAW could thus retain a certain proportion of its profits to invest in production, develop welfare programs or provide bonuses. Profit-sharing arrangements were further developed into what was known as a responsibility system in the early 1980s. Central government stipulated a baseline contribution by FAW of 137 million yuan in 1982, with an annual increase of 2.5% (FAW 2003b). FAW could keep all profits over and above this amount. In practice, from 1983 to 1986, profits retained by FAW reached about 400 million yuan, of which 230 million was re-invested in production and about 170 million was spent on employee welfare (FAW 2003b). The responsibility system was extended in 1985, when FAW was among the enterprises to which a separately listed account was assigned under the national planning system. Having this account meant that its production duties could be directly arranged by central government without any intermediate bureaucratic procedures. FAW was thus released from interventions by many branches of the state apparatus, and no longer had to negotiate its production plans with them.

As FAW gradually gained autonomy during the reform, the internal production arrangements were also re-organized and the previous hierarchical command structure abandoned. Encouraged by central government, responsibility systems were comprehensively adopted at different levels inside FAW in the 1980s. The production and management activities of the whole factory were accordingly divided into six categories, 24 schemes, and 240 goals, which were then distributed to workshops, work teams, and finally individuals. In total, approximately 250,000 specific production and/or administrative duties were assigned (see FAW 2003b). Each of these responsibilities was then converted into a 100-point score system that provided progress indicators for the whole process, and further allocated to a month-by-month work schedule. A timetable of responsibility was made available to each team or individual, and bonus and welfare arrangements were linked to the accomplishment of monthly tasks. Seven types of responsibility system were developed by FAW in the areas of profit, output, salary, loss, investment, single project, and "extra income". A few of these specific responsibility systems are worthy of elaboration. For example, salary responsibility stipulated that the total salary of FAW would be a fixed sum, regardless of increases or decreases in the number of workers. Similarly, the loss responsibility system was based on a designated sum, below which any reduction of loss would be rewarded. For the extra-income responsibility system, FAW's branch factories were allowed to earn additional monies through outside projects, so long as their production responsibilities were fulfilled, and 20–30% of this extra income could be retained. The seven responsibility systems were later expanded, forming a comprehensive management scheme of responsibility at FAW.

The process of increasing autonomy at FAW also entailed building a new governance structure between the state and the enterprise. On the one hand, FAW still followed directions from above to fulfill "national duties", regardless of its own business interests. For instance, throughout the 1980s, the central government decided to concentrate the industrial structure of the automobile sector. One of the major initiatives was to promote federations among large-scale automotive enterprises. and FAW, as the largest automaker, was thrust to center stage, incorporating many small automobile enterprises across the country, so that by the end of the 1980s, about 100 automobile enterprises together formed a production federation headed by FAW (FAW 2003b). This initiative was not good news for FAW, as it increased the administrative burden on the enterprise, but it was implemented nonetheless at the insistence of central officials.

On the other hand, the former mode of direct hierarchical state commands was no longer viable, especially when FAW came to realize its own interests in the market economy. This could not be more clearly evidenced than by the debate between the central government and FAW in 1984 about the newly established China National Automobile Industry Corporation (CNAIC). This corporation was meant to replace the old agencies in charge of the automobile sector within the Ministry of Machinery Industry, with the principal aim of integrating Chinese automobile enterprises toward a more concentrated industrial structure. However, when the CNAIC made an aggressive decision in 1984 to incorporate a few large automakers including FAW under its direct management, FAW managers immediately fought back. They appealed directly to central leaders to protect their newly gained autonomy, protesting that integration by CNAIC would drag FAW back into the planned-economy system, and "turn FAW into a big workshop for this corporation" (FAW

2003b: 261). The central leaders were persuaded, and the CNAIC was eventually mothballed by the state in 1987, to be replaced later by the Automobile Industry Federation which provided policy consultation and served as an intermediary between state agencies and automotive enterprises. This organization was finally renamed the China Association of Automobile Manufacturers in 1990.

The second wave of enterprise reform affecting FAW began in the early 1990s, when the central government determined to corporatize major enterprises and promote them into large business groups. Under the Regulation on Management Transformation of State-owned Industrial Enterprises, issued in 1992, state-owned enterprises would be transformed into independent judicial individuals in the market. The enterprises would hold management rights and take responsibility for gains and losses, while their assets would belong to the people of China, of whom the government is the ultimate representative. Under this structure, the state would recede from ordinary management of enterprises, but would practice macro-management and monitor them for the sake of state-owned assets. As well as the objective of corporatization, central leaders also wished to promote large-scale business groups, in order to address the scattered industrial structures inherited from the planned-economy system. Policymakers believed that these groups could achieve greater economies of scale, promote organizational structures, encourage intra-group coordination, and increase competitiveness, taking it for granted that significant enterprises such as FAW would serve as the platform for organizing these business groups.

Such were the new policy orientations that paved the way for the organizational transformation of FAW in the 1990s from a state factory into a modern corporation at the head of a large business group. In 1992, FAW was asked to trial organizing the business group, and as the China FAW Group Corporation, became the core organization of the business group, in which a total of 152 member enterprises were to be incorporated. A more detailed plan was developed by FAW managers, in which the FAW group would be structured into three levels: the core level would comprise 37 specialized sub-factories, three wholly owned subsidiary companies, and three directly affiliated research institutes; the closely associated level consisted of FAW-held companies and jointly held companies; and the loosely associated level was for corporations coordinating with FAW. According to the plan, all factories at the core level would be gradually converted to sub- or subsidiary companies; and necessary institutions for corporate governance would be set up for the FAW group, such as a board of directors and general manager responsibility systems (Gen 1993).

As early as the 1980s, FAW had been allowed to be an experimental enterprise, entitled to manage state-owned capital. In 1993, FAW was officially authorized to manage all of the state-owned assets in the group, and encouraged to consolidate the business group it had been building up via capital operations. Incorporating other enterprises in the market had never been easy, as it involved complicated procedures of re-arranging ownership and management between different local government units, as well as within different branches of central government. Very often, a central government order was not actually sufficient for FAW to acquire an enterprise, and FAW had to compensate stakeholders, who were mostly local government actors. However, with its official authorization to manage state capital, the business group established by FAW was no longer a loose federation of associated enterprises as in the 1980s, but built on a solid ownership structure, the mainstay of any business group in the modern sense. By 2001, FAW had entirely taken over 38 domestic enterprises, 12 of which were converted to sub-companies or wholly owned subsidiaries, 11 as holding companies, nine as jointly held companies, and six as directly affiliated sub-factories (Historical Record Office of First Auto Works 2002).

As part of the corporatization movement among large enterprises initiated by the state, FAW started to adjust its enterprise structure by detaching directly affiliated sub-factories. Holding the entire production procedure in a single large enterprise was a historical legacy of China's automobile sector, traceable to the replication of the Soviet production systems. However, in the process of corporatization, such a production system was seen as a barrier to better specialization within the group. De-affiliation initially started with the auto-parts production segment. Out of nine auto-parts plants within the corporate group, a single auto-parts corporation, named Fu-Ao, was formed in 1998. This move stripped about 15,000 employees from the overall labor force of the group. Later, more plants were discharged from the core as subordinate corporations. In 1999, based on the previous sub-branches of FAW, four functional companies were established. The next year, another five sub-factories were separated from the core corporation and converted into wholly owned subsidiaries or holding corporations under the FAW group.

A further important step towards corporatization was to develop a modern system for corporate governance within the group. In 2001, the

FAW group applied to the central government for permission to set up a wholly state-owned corporation as its core corporate structure, with a board of directors in charge of strategic decisions and a management committee responsible for ordinary management. In the same year, FAW issued a series of governance principles regulating the relationship between its core and subordinate enterprises. These principles matched well with the standards of a modern business group, which stressed autonomy of group subordinates under the macro-direction of the core.

After all these rounds of reform, FAW had finally matured as a large-scale business group equipped with modern corporate institutions. By 2001, the group comprised one core corporation wholly owned by the state, nine directly affiliated sub-factories, 23 functional departments, 25 wholly owned subsidiaries, 13 holding companies, 30 jointly held companies, and 224 coordinating enterprises (Historical Record Office of First Auto Works 2002). As a "central enterprise", FAW is subject to SASAC, the state agency which monitors state-owned assets, and its top director should be appointed by the Politburo of the Central Committee of the CPC. Although FAW behaves like a normal market player, in fact it serves the central government as a part of its controlling apparatus over the national economy.

Although the local governments in Changchun and Jilin still actively assisted with the organizational transformation and technological upgrading of FAW post reform, their relationship with FAW was based more on the interests and concerns of the local economy than on bureaucratic orders as in the old days. Through the course of China's market reform, FAW had become a vital economic asset driving local growth. Despite a proportion of its profits being earmarked for central government, FAW still had a significant influence on the local economy, not least through its huge demand for local labor, materials, and energy, as well as its production chain nurturing numerous local automotive enterprises. From 1953 to 1983, FAW developed 5 local automobile and motor assemblers, 11 local retrofitting factories, 38 local accessory factories, and 88 auto-parts factories for Jilin Province (Zhao 1983). For this reason, reform-era economic plans from Jilin Province and Changchun city always mention the automobile sector as a major feature of local comparative advantage. In the minds of local officials, the automobile sector was a pillar of the local economy, and FAW was the core of this pillar.

Local governments, either municipal or provincial, were therefore always willing to coordinate with central government to transform and upgrade FAW after the reform, especially as regards infrastructure construction. For example, when car production loomed large on FAW's agenda in 1986, the municipal government of Changchun confiscated a wide area of land covering 535 hectares for a new production base, evicting about 10,000 peasants in the process. In the same year, when the truck remodeling project was at its most critical stage, provincial leaders paid multiple visits to FAW to provide support. In 1991, when the joint venture was proposed, FAW-VW quickly became the number one project for Jilin Province, and local political leaders organized a headquarters with FAW directors to coordinate all related construction activities (FAW 2003b). Reflecting on decades of close cooperation and intensive assistance from local government, at FAW's fifty-year anniversary ceremony in 2003, provincial leader Wang Yunkun declared the achievements of FAW not only a success for the enterprise, but the pride of all the people of Jilin (Archive of First Auto Works 2004).

Technological Upgrading via Foreign Partners

The central government also directed FAW's moves towards technological upgrading during the reform. Clearly aware of the backward state of China's automobile industry, national policymakers sought to boost its technical proficiency and capabilities. As with all other aspects of the reform process, FAW became the standard-bearer of this national mission, and technological upgrading began with the self-reliant remodeling of its long-manufactured Jiefang truck. However, when national requirements dictated a shift of focus to the much more challenging field of passenger cars, central leaders acknowledged the need to open industry access to foreign corporations in order to introduce advanced models and related manufacturing technologies. Joint ventures with foreign automakers thus became an official strategy for the technological upgrading of Chinese car production, with FAW as one of the major practitioners.

When China started to open its gates as the market reform progressed, central officials were astonished by the serious underdevelopment of the Chinese automobile industry in comparison with world standards. In 1978, central administrators sent a study team of about twenty FAW managers and technicians to Japan to make contact with and draw lessons from the significantly more advanced Japanese automakers. During the six-month tour the team—divided into four groups of product design, techniques and quality, production assistance, and enterprise management—first spent a month visiting the ten top Japanese automakers, including Toyota, Nissan, and Honda, then undertook immersive studies for one month at each of Mitsubishi, Suzuki, Nissan, Hino, and Toyota. Findings from the Japan tour were summarized in 26 books of study materials (Sheng 1998).

The replication of the Stalin Automobile Factory in the Soviet Union had been the first stage for the nascent Chinese automakers to learn how to produce vehicles using external expertise. Decades later, the trip to Japan was only the second such opportunity, and it opened the eyes of the Chinese industry to a new world (Xu 1983). This time, a massive technological gap was quickly perceived. In particular, due to the lean production model of Japanese automakers, the labor productivity of Toyota was 70 times of that of FAW. Had the lean production model, however desirable, been applied in China at that point, FAW would have had to lay off an unacceptable number of workers. Central administrators soon issued an urgent call, keenly echoed by FAW, to catch up with the advanced foreign automakers. Being more autonomous in the new economic environment, FAW also became eager to upgrade its old technology to compete for a large share of the domestic market. Consequently, from the early 1980s the technological upgrading of FAW progressed in a very aggressive manner. Table 2.3 provides a detailed comparison between Toyota and FAW in 1978, summarized by FAW engineers on the tour in Japan.

The central mission of this movement towards technological upgrading from 1980 to 1986, often termed "the remodeling project" at FAW, was to remodel the old Jiefang truck. The modest target for the new model was to approach the international standards of the late 1970s and early 1980s. The project, unprecedented in terms of invested resources, resulted in a new model of 5-ton truck, the CA141. According to a thorough appraisal by Japanese automakers Hino, the CA141 was a technological advance on FAW's CA15 (the latest model of the CA10 family), but remained behind the Hino GD174 truck model in the 1980s (FAW 2003b). Although still not as advanced as the Japanese products, the CA141 was a successful model, which indicated that the technology and production capabilities of FAW had moved into a new age.

As stipulated by the state, the R&D of the new CA141 model was self-reliant both financially and technically. During the planned-economy era, all necessary investments to develop the new model would have been allocated by central government. However, this time, central administrators provided no financial endorsement, merely general direction, a pattern that was consistent with the new state–enterprise relationship following the reform. Any assistance provided by the government was usually market oriented. For example, by allowing FAW to retain a certain degree of

Table 2.3 Comparison between Toyota and FAW in the late 1970s

Items	*Toyota*	*FAW*
Product development		
Time to develop a new model	3 years	8–10 years
Time for new model to achieve normal productivity	4 days	1–1.5 year
Time for new model to achieve normal quality	1.4 months	1–1.5 years
Sales and production		
Sales cycle of new product	3 days	>3 months
Ratio of storage time versus production time for components	300–500	5000–10,000
Organization and technology in production		
Annual output per capita	57.7 units	1.0–1.5 units
Time to change production mold	3 minutes	0.3–2.0 hours
Time of material reservation	3 days	0.5–1.5 months
Labor organization and skills		
Ratio of mechanical operation per capita	95%	29%
Labor productivity with FAW as reference	70	1
Coordination and logistics		
Coordination between assembler and auto-parts producer	70–80%	30%
Cost of logistics with FAW as reference	1	>10
Quality management		
Overhaul distance in kilometers	>300,000	100,000
Equipment management		
Ratio of equipment activation	100%	70%
Computers and information		
Ratio of computer application	>90%	30%
Cost control and financial management		
Profit ratio of cost	>20%	<15%
Ratio of capital occupation	1%	25%

Source: Based on Chen (1998)

profit, the state enabled the enterprise to accumulate the necessary investment capital for the remodeling project; when these self-accumulated funds were not enough, the state offered financial assistance in the form of loans. Moreover, the remodeling project was mainly accomplished by the R&D team from FAW. Originally, there had been proposals from within FAW advocating the introduction of upgraded technologies by foreign corporations, but this would have come at unacceptably high costs. In comparison with the state-led technology transfer from the Soviet Union in the early days of FAW, the official rejection of the proposals had the effect of pushing the remodeling project into a self-reliant pathway. In spite of FAW's limited R&D experience, the design of the new product was accomplished quite quickly: in 1980, R&D for a new truck model and corresponding engine was assigned to the R&D department of FAW; a prototype model was built one year later, and approved by the central government by 1983.

When automobile production began to concentrate on passenger cars in the late 1980s, however, setting up joint ventures with foreign partners became the main solution for technological upgrading. As the chosen industry leader in the duty of import substitution, FAW was strongly encouraged by the state to set up joint ventures, introducing foreign technology and promoting domestic production. Opening to foreign capital and technology was gradually gaining political legitimacy among the top politicians of China after the market reform. In practice, introducing foreign technology was also more realistic for passenger cars: car manufacturing technology was much more sophisticated than the truck remodeling project, and none of the Chinese automakers of the 1980s had ever produced modern cars on a large scale.

With the support of the state, FAW was careful in its selection of potential foreign partners for joint ventures. The three foreign automakers with the greatest potential in the eyes of FAW were Chrysler, Volkswagen, and Mitsubishi. FAW initially showed great interest in forging a connection with Chrysler in the United States, seeking to introduce its Dodge 600 model, including engine technology and body design. In 1987, after several rounds of negotiations, FAW and Chrysler signed a contract to introduce the American engine technology, through which FAW could acquire an entire assembly line for the engine with all necessary technological documents. However, when negotiations proceeded to the car body, Chrysler set an extraordinarily high bid. From a business perspective, this was a common strategy, since Chrysler already held an advantage with the engine contract. However, the FAW leadership, who lacked market negotiation experience, regarded the offer as "unacceptable" and "ridiculous" (Li 2001: 3). To make matters worse, no major directors of Chrysler appeared at the new negotiations, at which the Chinese participants took offence. Chrysler immediately lost the trust of FAW and was no longer considered a potential partner for joint ventures.

With the FAW–Chrysler cooperation sailing into troubled waters, Volkswagen (VW) seized the opportunity to step in. Keen to gain access to the Chinese market, VW made a very generous offer to FAW: it proposed introducing its top model, the Audi 100, promising to install all the required technology and realize domestic production. Under the deal, VW would help FAW transform its old Hongqi model by fitting the already introduced Chrysler engine into the Audi 100. This offer would ensure the Chrysler engine would not be wasted, and equally importantly, the use of Audi technology under the old Hongqi brand was politically acceptable. For FAW, VW was also much more sincere in its negotiations than Chrysler. With the initial proposal on the table, the top VW director visited Changchun to meet FAW's directors. In 1988, a formal contract was signed between FAW and VW, a partnership which has lasted until the present day. The third player, Mitsubishi, briefly contacted FAW in November 1987 to explore the potential for cooperation, but FAW soon realized that the model they proposed was outdated, and concluded that Mitsubishi, like Chrysler, did not take the Chinese market seriously (Zheng 1998).

The deal between FAW and VW proved a major success. As an initial project, the new Hongqi CA7200 model developed by VW included a high proportion of domestic content, and led to the establishment of the FAW-VW joint venture in 1991, in which FAW held a 60% stake and VW 40%. Two popular VW models, the Jetta and Golf, were subsequently introduced through the joint venture; and a luxury model, the Audi CSV6, was introduced to the production portfolio in 1995. FAW-VW became a focal project in the Chinese automobile sector, with strong support from the central government. In 1991, the National Planning Commission organized a conference in Changchun for the construction of necessary additional facilities for FAW-VW. The following year, the Ministry of Machinery Industry held two consecutive meetings at FAW, coordinating production of the equipment needed for FAW-VW; and in 1994, the National Planning Commission organized a further meeting in Beijing to accelerate the construction of the FAW-VW plant. By 1996, the construction of FAW-VW was formally accomplished, and it became the largest joint-venture project in China's automobile industry: the total investment reached 10 billion yuan, with an annual output of 150,000 units of popular models and 30,000 units of luxury models. The Jetta reached domestic coverage as high as 84%, with the Audi approaching 60% (FAW 2003a), dominating the Chinese domestic market in the late 1990s and early 2000s.

In the new century, FAW strategically selected Toyota as its second foreign partner. In the 1980s, Toyota had been hesitant about moving into China. However, as the Chinese market grew rapidly in the 1990s, the directors of Toyota changed their position and started to make moves towards opening operations in China. Its first trial, established in 1998, was a joint venture in Sichuan Province to introduce a minibus model, the Toyota Coaster, rather than cars. After the trial, Toyota transferred its attention to Tianjin Automotive Industry Corporation (Tianjin Auto), which was one of the "Small Three". Tianjin Auto had participated in a long-term cooperation with Japanese car firm Daihatsu since the early 1980s, as Toyota was well aware. In 1998, when Toyota was ready to expand in China, it took over Daihatsu, setting up a joint venture with Tianjin Auto in 2000. However, as Tianjin Auto could not fully satisfy the production requirements of Toyota, FAW became its next target. Cooperation with Toyota was also an important strategic choice for FAW. As one of the world's largest automobile corporations with cutting-edge production technology, Toyota was an ideal partner. Moreover, having two foreign partners rather than one could greatly increase FAW's bargaining power in future business negotiations. In order to establish cooperation with Toyota, FAW first incorporated Toyota's joint ventures into its group in 2002; then FAW and Toyota announced their comprehensive cooperation. The agreement with Toyota finalized FAW's grand joint-venture strategy.

To date, the ongoing partnerships with VW and Toyota remain FAW's two major platforms in terms of technological upgrading. In recent years, the FAW-VW and FAW-Toyota joint ventures have consistently been among the national top ten automobile manufacturers by sales. FAW-VW reached accumulated production of 5 million vehicles by 2011, and FAW-Toyota reached the same target by 2016. The significant production of these joint ventures keeps FAW as one of the largest carmakers in China, second only to SAIC (see Chapter 3).

Case Summary

Located in northeast China, FAW is the country's oldest automobile manufacturer, and represents the first production model, in which an industrial enterprise is owned by the central government and focused on producing foreign-model cars via joint ventures. The story of FAW empirically validates the institutionalist framework proposed in this book. With this framework in mind, the origins of the FAW production model are traceable to the national positioning of the northeast in the pre-reform era. The destiny of FAW in the market transition was deeply associated with the decision of communist China's central leaders to build the northeast region as the country's national heavy-industry base, with FAW as the core national automobile production enterprise. When the reform started in the 1980s, this prior economic positioning had a strong impact in shaping the ownership structure and technological strategy of FAW. This institutional path dependency was manifested in the changing local political structure and configurations of developmental ideas surrounding FAW: considered a reliable channel through which the state to could leverage the new market environment, the central government maintained a firm hold on FAW in the reform, while reorganizing an outmoded factory and management structure to become a state-owned corporation. As the "senior son" of the automobile sector, FAW was appealed to by the state to trial the unparalleled developmental approach of introducing foreign technology within a state enterprise, via joint ventures with foreign corporations. The resulting developmental trajectory was an evolutionary sociopolitical process consistently led by central government, which placed FAW at the forefront of different phases of state enterprise restructure, and used FAW as the main means to enact the new national industrial policy direction. The analyses in this chapter reveal the origins of the distinct ownership structure and technological strategy of FAW, and the first production model under study, in the market transition.

Notes

1. "156 projects" were a large-scale technology transfer from the Soviet Union to China in the 1950s and 1960s resulting from the post-World War II political alliance between the two countries, with the main purpose of assisting the development of China's heavy industry. This transfer took multiple forms, including introducing equipment, providing technical drawings and documents, sending Soviet experts to China, training Chinese technicians, and establishing engineering schools and research institutes in China. Although the Soviet Union had suspended this program by early 1960s as Sino-Soviet relations deteriorated, these projects greatly facilitated development of the Chinese heavy sectors (Zhang et al. 2004).

Bibliography

Archive of First Auto Works. (2004). *FAW Yearbook 2003*. Changchun: First Auto Works. (in Chinese).

Central Committee of the Communist Party of China. (1999). The Decision of the Central Committee of the Communist Party of China on the Reform and

Development of the State-owned Enterprises. *Announcement at the Fourth Plenary Session of the 15th Central Committee of the CPC.* (in Chinese).

Chen, J. (1998). The Analyses and Suggestions for First Auto Works to Introduce the Lean Production. In Historical Record Office of First Auto Works (Ed.), *Studying in Japan* (pp. 198–241). Changchun: First Auto Works. (in Chinese).

Dong, Z. (2004). Examination of Old Industrial Base Invigoration Based on Characteristics of Infrastructure Investment of the Northeast in the Latter Half of the Twentieth Century. *Study of History of the Chinese Communist Party, 5*, 50–57. (in Chinese).

First Auto Works. (2003a). *First Auto Works Fifth Years Memorabilia, 1953 to 2003.* Changchun: First Auto Works. (in Chinese).

First Auto Works. (2003b). *The Path of First Auto Works in Fifty Years, 1953 to 2003.* Changchun: First Auto Works. (in Chinese).

Fu, Y. (2004). The Strategy and Construction of the Central Government in Northeast China in 1950s. *History Study of the Chinese Communist Party, 5*, 58–64. (in Chinese).

Gen, S. (1993). Push Reform, Enlarge Opening, Catch Opportunity, and Make Accomplishment. In Historical Record Office of First Auto Works (Ed.), *First Auto Works Yearbook 1993* (pp. 29–40). Changchun: Jilin Science and Technology Press. (in Chinese).

Historical Record Office of First Auto Works. (1991). *The Factory History of Fist Auto Work.* Changchun: Jilin Science and Technology Press. (in Chinese).

Historical Record Office of First Auto Works. (1998). *The History of Hongqi Car.* Changchun: First Auto Works. (in Chinese).

Historical Record Office of First Auto Works. (2002). *FAW Yearbook 2001.* Changchun: First Auto Works. (in Chinese).

Li, G. (1983). Four Important Steps to Revitalize the Chinese Automotive Industry. In China National Automotive Industry Corporation (Ed.), *China Automotive Industry Yearbook 1983* (pp. 40–50). Beijing: Machinery Industry Press. (in Chinese).

Li, H. (1996). The Basic Ideas of Strategically Restructuring the State-Owned Economy in the Ninth Five-Year Plan. *China Industrial Economy, 7*, 5–8. (in Chinese).

Li, A. (1998). *Family Cars Attracting China.* Beijing: Writers Press. (in Chinese).

Li, G. (2001). Reliance or Self-Reliance, Chinese Automobile Industry in the Opened Environment. *Auto & Society, 4*, 8–11. (in Chinese).

Ma, J., & Liu, H. (2000). *The History and Perspective of the Reform of the State-Owned Enterprises in China.* Beijing: Capital University of Economics and Business Press. (in Chinese).

National Statistics Bureau. (1999). *Major Structure Adjustment and Rapid Economic Growth.* http://www.stats.gov.cn/ztjc/ztfx/xzg50nxlfxbg/200206/t20020605_35960.html. Accessed 10 June 2017. (in Chinese).

Rao, B. (1983). United Together, Learn from Experience, and Push the Modernization of the Automobile Industry. In First Auto Works Research Center (Ed.), *The Cradle of the Chinese Automobile Industry* (pp. 22–30). Frist Auto Works: Changchun. (in Chinese).

Rao, B. (1984). Foreword. In China National Automotive Industry Corporation (Ed.), *China Automotive Industry Yearbook 1983* (pp. xi–xiii). Beijing: Machinery Industry Press. (in Chinese).

Sheng, J. (1998). Prelude of the Remodeling Project: Memory of 160 Days in Japan. In Historical Record Office of First Auto Works (Ed.), *Studying in Japan* (pp. 1–14). Changchun: First Auto Works. (in Chinese).

Sino-Soviet Friendship Association in Jilin and Changchun. (1956). *First Auto Works is Fruit of Friendship between Chinese and Soviet People*. Changchun: Jilin People's Press. (in Chinese).

State Council of China. (1994a). *Outlines of the Industry Policy in the 1990s*. State Council of China. (in Chinese).

State Council of China. (1994b). *The Industry Policy of the Automobile Industry*. State Council of China. (in Chinese).

Sturgeon, T., & Florida, R. (2000). *Globalization and Jobs in the Automotive Industry*. Industrial Performance Center, Massachusetts Institute of Technology, Working paper Series. https://ipc.mit.edu/sites/default/files/documents/00-012.pdf. Accessed 10 June 2017.

Wang, X. (1998). From 'Specimen' in the Planned Economy to 'Red Flag' in the Market Economy. *Outlook, 17*, 30–32. (in Chinese).

Xu, Y. (1983). Reflection and Prospect. In First Auto Works Research Center (Ed.), *The Cradle of the Chinese Automobile Industry* (pp. 1–13). Changchun: First Auto Works. (in Chinese).

Yang, W. (2005). The Historical Contribution of the Northeast Old Industrial Base to the New China. *References of Economic Study, 91*, 24–28. (in Chinese).

Yi, B., & Lin, S. (2001). The Industrialization History of the Northeast. *Northeast Asia Forum, 4*, 54–64. (in Chinese).

Zhang, B., Yao, F., Zhang, J., & Jiang, L. (2004). *The Technological Transfer from Soviet Union to China, 1949 to 1966*. Jinan: Shandong Education Press. (in Chinese).

Zhao, X. (1983). Concentrate to Help the Remodeling in First Auto Works. In First Auto Works Research Center (Ed.), *The Cradle of the Chinese Automobile Industry* (pp. 14–16). Changchun: First Auto Works. (in Chinese).

Zheng, Y. (1994, August 9). Critique on the Car Civilization. *Guangming Daily*. (in Chinese).

Zheng, Y. (1996). *Debate on Cars*. Beijing: Economic Science Press. (in Chinese).

Zheng, Y. (1998). My Memory on the Project of 30 Thousand Cars in First Auto Works. In Historical Record Office of First Auto Work (Ed.), *The History of Hongqi Car* (pp. 212–227). First Auto Works: Changchun. (in Chinese).

Zheng, M. (2000). The Expansion of Japanese Capital in the Occupied Northeast. *The Frontline of Social Science, 6*, 183–190. (in Chinese).

CHAPTER 3

SAIC, a Giant of the Domestic Market

The Shanghai Automotive Industry (Group) Corporation (SAIC) is currently China's largest manufacturer of passenger cars. If First Auto Works (FAW) is considered the "senior son" of China's planned-economy era automobile industry, it is fair to describe SAIC as opening the gate to modern car manufacturing in China during the market reform. Unlike the story of FAW, which was primarily orchestrated by the central government, SAIC was owned and directed by the Shanghai municipal government, with the concrete support of the central authority. One of SAIC's most telling undertakings was to pilot the establishment of a joint venture with Volkswagen, Shanghai-VW. While joint ventures later became both common and highly successful, as described in Chapter 2, the initial move by SAIC broke new ground for the Chinese automobile sector in terms of how another production model might emerge, characterized in this case by local government ownership and close cooperation with foreign partners. The joint-venture strategy would later prove influential over the direction of travel taken by centrally managed enterprises, such as FAW, and the Chinese automobile industry as a whole.

This chapter begins with an overview of Shanghai's historically central position in the Chinese economy and the significance of this legacy after the economic reform. As the constraints of the planned economy were gradually lifted in the early stages of market reform, local government in Shanghai strengthened throughout the 1980s and 1990s. In particular, the extraordinary power of Shanghai's political leaders, who often later rose to high office in the central government, made them wield a degree

Q. Feng, *Variety of Development*,
https://doi.org/10.1007/978-981-10-5912-4_3

of influence over and leverage support from the central level when needed. The chapter next explores how local officials developed Shanghai's automobile sector with the assistance of foreign carmakers, as a way of catalyzing technological upgrading, and how these ideas were eventually accepted and supported by central government. Finally, the chapter charts the socio-political process of SAIC as a corporation under local government ownership and direction, with central support, through engaging foreign joint-venture partners.

Shanghai, China's Economic Capital

In comparable ways to the historic significance of the northeast to Chinese industry, Shanghai had been a major economic center long before the founding of the People's Republic of China, and its position within the planned economy therefore became a core concern for the communist leaders. Located at the mid-point of China's eastern seaboard, the city of Shanghai had a glorious past, rising to national prominence as a financial, commercial, and industrial center over the course of just under a century, from the first Opium War in the early 1840s to the Sino-Japanese war in the late 1930s. From the 1870s to the 1930s, Shanghai's annual foreign trade often accounted for about half of the national total; and by the 1930s, before the second Sino-Japanese war, one-third of all national commercial wholesalers were based at Shanghai, selling 60% of their goods to other domestic markets. Over the same period, 58 of China's 164 banks were headquartered in Shanghai, and members of the Shanghai Bankers' Association held about 90% of total national financial assets. Local industry held 40% of total national industrial assets and employed about 40% of all industrial workers in China (Jin 1984; Committee of Shanghai Local Record 2005). As the noted Chinese economist Ma Yinchu commented, "though not as advanced as London in Britain and New York in United States, with regard to the economic importance for the nation, Shanghai is truly comparable" (see Ma 1988: 79).

Such a city, in the minds of national policymakers, deserved special consideration as to how it could be positioned to serve the national economy. When Shanghai was taken over during the civil war in 1949, official expectations of leading post-war economic reinvigoration immediately surrounded the city. After the communist victory, Shanghai was accorded municipality status directly under the central government, the top level of administration equivalent to that of a province. Chen Yi, one of the top political leaders in the Communist Party of China (CPC), was carefully chosen as the first mayor of the new municipality. However, the initiative to tune up this potential economic engine for the new China was put on hold from 1949 to 1955, when Shanghai was considered vulnerable to coastal military threats from the Kuomintang government, which by then was retreating to Taiwan. Among the "156 projects" of the first national five-year plan (1953–1957), Shanghai received nothing. When military concerns began to subside, however, central leaders soon revived their enthusiasm towards the city. Re-building Shanghai to better serve the Chinese economy was declared a national mission. In 1956, Chairman Mao issued his well-known edict directing socialist construction in China, *On the Ten Vital Relations*, in which he proposed to take advantage of the existing industrial foundation in the coastal areas. In the same year, the party leader specifically emphasized that the development of Shanghai should be placed high on the state agenda (Tang 2001). Echoing Mao's proclamations, the Shanghai government proposed a developmental principle for the local economy that emphasized harnessing the industrial potential of Shanghai, and properly developing local industrial production accordingly. This guideline was further specified at the first municipal congress of the CPC in Shanghai:

> ...the central point is to fully use the historical industrial foundations in Shanghai and apply its previous equipment and technology to better meet the needs of the country and people, and accelerate the national socialist construction. (Ke 1956: 576)

In accordance with the dictates of central planners, the economy of Shanghai was comprehensively reshaped in the following ways. First, local private enterprises predominant in the pre-communist era were mostly transformed to state- or collectively owned enterprises. The shift of ownership structure in Shanghai was part of a national political and social movement launched in 1956, aimed at nationwide transition to socialist public ownership. Shanghai, hosting the largest capitalist economy in China, was an obvious major target of this movement. Soon, private industrial enterprises in Shanghai disappeared (Committee of Shanghai Local Record 2005): the number of non-state- or collectively owned industrial enterprises plummeted from 22,977 in 1955 to only 6 in 1956, while state-owned industrial enterprises concurrently soared from 292 to 17,090 over the same period. The market share of private retailers similarly fell from 91.6% in 1950 to just 1.8% in 1956, by which time private wholesalers had completely disappeared. The same year, formerly prosperous privately run financial institutions were mostly dismantled, with only 3 survivors out of the 167 private banks or money exchanges that had populated the local financial sector in 1949.

Second, the economic functions of Shanghai radically changed to match the national designation, which saw some of the city's former economic roles all but erased with the establishment of the planned economy. Perhaps the most striking aspect of this shift was Shanghai's shrinking prominence as an international trade hub, as China became increasingly isolated from the Western world in the 1950s. Beneath the surface, however, the state-led socialist transformation further transferred control over commodity distribution and financial circulation into the hands of officials, meaning that Shanghai would also no longer act as the national hub for domestic trade and financial markets. Instead, in accordance with the new developmental principles set out above, the main function of Shanghai in the planned economy would be to focus on industrial production to serve the national economy.

Third, the central government directly modified the industrial structure of Shanghai in line with national planning priorities. As the former economic center of pre-communist China, Shanghai had long concentrated on light industry, with major sectors such as textiles, cigarettes, matches, soap, flour, leather, and rubber accounting for about three-quarters of local industrial production output value. By comparison, local heavy industry was rather underdeveloped: equipment manufacture comprised 8.5% of total output value, and raw materials such as steel and chemicals accounted for only about 3% (Committee of Shanghai Local Record 2005) However, as the national planning focus pivoted towards heavy industry in the 1950s, the direction of the Shanghai economy soon followed suit. In particular, as the city's existing light industry was considered to provide a relatively good industrial foundation, the central government compelled Shanghai to promote advanced industrial technologies and products to serve the needs of the new China. This policy paradigm is well evidenced in the early five-year plans in Shanghai. The second five-year plan (1958–1962), for example, stated that:

> We should transform and develop Shanghai in the following three directions: firstly, to build Shanghai as an industrial city, which is able to manufacture advanced light products of more varieties and with less consumption

> of raw materials; secondly, to build Shanghai as a national industrial base for new technologies and new products; thirdly, to maintain and develop coordination with local and national industry, and to make the machinery and electronics industries in Shanghai the supply base for non-standard equipment and accessories. (Planning Committee of Shanghai 1957)

The economy of Shanghai was thus systematically modified to match the national blueprint set out by the central planners, realigning the local industrial structure in favor of the prioritized sectors and the promotion of technological advances. As illustrated in Fig. 3.1, from 1952 to 1960, the output of the tertiary industries on which Shanghai's historic prosperity had been founded, including trade and finance, halved from about 40% to 20%. Over the same period, the output of non-construction secondary industries expanded from 50% to approximately 75%. Within the secondary industries, heavy-sector output amounted to about one-seventh of the output of light industry in the early 1950s, but was about 30% higher than light industry by 1960. This sector composition was maintained until the end of the planned economy era in 1978, and contributed to Shanghai's consolidation as an important center for both light and heavy sectors in

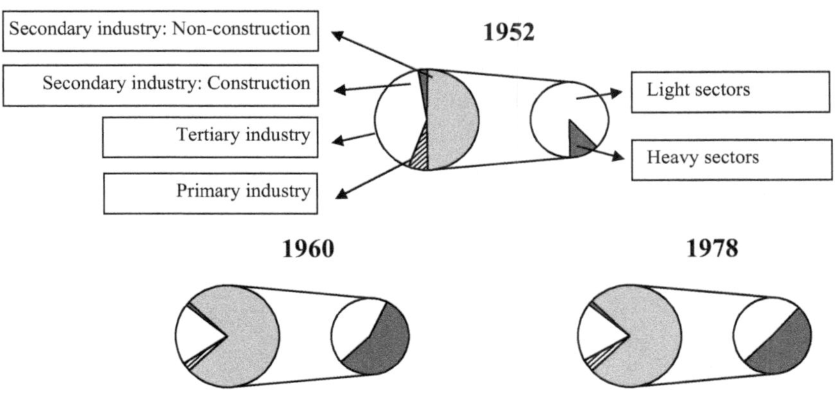

Fig. 3.1 Industrial structure of Shanghai, 1952–1978
Note: Under each year, the left pie represents proportion of output value by primary, secondary and tertiary sectors (secondary industry is further divided by construction and non-construction sectors), while the right pie shows the proportion of output value of the non-construction light and heavy sectors
Source: Committee of Shanghai Local Record 2005

communist China. According to 1984 statistics, the municipality produced about 13% of China's total steel, 27% of electricity-generating equipment, 10% of mining equipment, and 13% of metal-cutting machine tools; at the same time, Shanghai's light industry output was 12% of the national total, accounting for 20% of bicycle production, 30% of sewing machines and watches, 20% of televisions and tape recorders, and about 35% of cameras (Cao et al. 1987).

As industry expanded through the planned economy era, so Shanghai continued to grow in importance to the function of the Chinese economy, becoming a major cash cow for the central government with a critical role in the national account. From the 1960s to the end of the planned-economy period in 1978, Shanghai's contribution to national fiscal income maintained a consistent average of about 15%, while accounting for only about 1–3% of national expenditure (Table 3.1). The extent of this imbalance indicates great fiscal outflows from Shanghai to the central government from 1950 to 1980, the fiscal contribution from Shanghai to the national account reached almost 200 billion yuan, about 85% of Shanghai's total local fiscal income over the same period (Yao 1994). Between 1960 and 1980, there were 9 years in which Shanghai's annual turn-in ratio exceeded 90% of local receipts (Committee of Shanghai Fiscal and Taxation Record 1995).

The vital position of Shanghai in the planned economy had important implications for the relationship between local political leaders and the central authority. The key positions in the Shanghai municipal government,

Table 3.1 Fiscal income and expense of Shanghai in the national account (unit: 1 billion yuan)

Year	*Shanghai fiscal income*	*National fiscal income*	*Local/ national ratio*	*Shanghai fiscal expenditure*	*National fiscal expenditure*	*Local/ national ratio*
1950	0.08	6.52	0.01	0.08	6.81	0.01
1955	0.21	27.20	0.01	0.19	26.93	0.01
1960	10.01	57.22	0.17	1.94	65.41	0.03
1965	6.18	47.33	0.13	0.78	46.63	0.02
1970	9.84	66.29	0.15	1.29	64.94	0.02
1975	13.42	81.56	0.16	2.65	82.09	0.03
1978	16.69	112.11	0.15	2.60	111.10	0.02

Source: Zhu and Gan et al. (1999)

the mayor and the party secretary, were always determined by the senior national leadership; selected from among the political elites, they were usually more influential and promising than those appointed to other cities and provinces. It was also expected that those selected as Shanghai leaders would rise to high positions in the central government after a few terms in office. In consequence, the local government of Shanghai not only exhibited great loyalty, observing bureaucratic orders in the national interest, but also wielded extraordinary bargaining clout with the central administration when local interests required it. In this sense, the local government of Shanghai was relatively powerful in political leverage by comparison with many other local authorities in China.

Ups and Downs in Reform

The development trajectory of Shanghai, which was somewhat skewed in the economic reform process compared to other administrative regions, can be divided into two stages. In the 1980s, as market relations began to transform the Chinese economy elsewhere, Shanghai stagnated, as central leaders sought to buffer the critical economic position of Shanghai as a safeguard against the major uncertainties and risks associated with the reform. By the 1990s, as confidence grew, this administrative barrier was lifted and Shanghai was soon revitalized by a strong and active local developmental state that became a leading force in the accelerating market economy. These dynamics of the local economy and associated changes in the local political structure provide clues to the origin of the production model later pioneered by SAIC.

The Lost Decade

The 1980s was a lost decade for the Shanghai economy compared with what was happening elsewhere in China. As the market transition took hold, Shanghai's historically leading contribution to the national economy began to decline. From 1980 to 1990, annual GDP growth in Shanghai fell mostly below the national average (Fig. 3.2). In terms of national ranking, Shanghai's local GDP growth rate slipped from among the top three performers to 24th, while per-capita income growth fell from the national top three to 25th (Shanghai Zhengda Institute 2002). Shanghai's stagnation appeared ever more evident when compared with other regions, which were experiencing rapid economic transformation in China's new

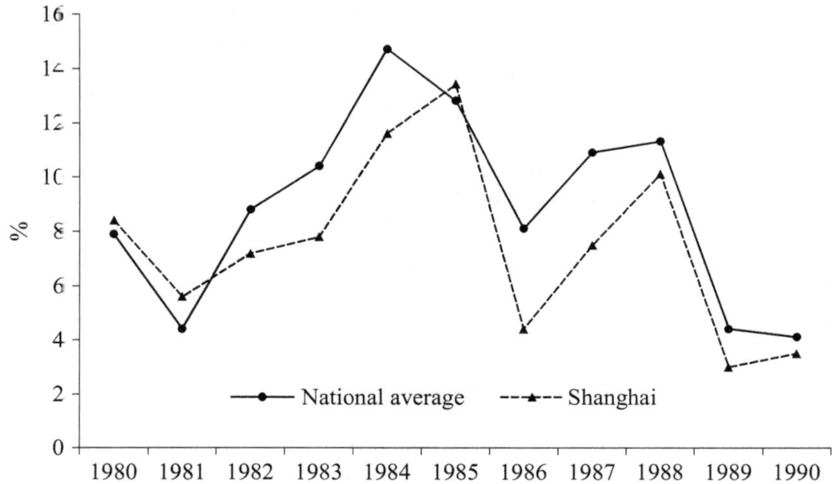

Fig. 3.2 GDP growth rates from 1980 to 1990, Shanghai and national average
Source: Based on Yao 1994

market environment. Throughout the 1980s, average annual GDP growth in the four coastal provinces of Guangdong, Fujian, Zhejiang, and Jiangsu reached as high as 10.5%, 12.6%, 11.8%, and 10.3% respectively, while the average rate for Shanghai was only 7.5% (Shanghai Zhengda Institute 2002), resulting in a marked decline in the economic gravity of Shanghai within China as a whole. In 1978, Shanghai accounted for 7.6% of national GDP, while local industrial output was 13.0% of national output, but these values had reduced to 4.2% and 4.9% respectively by 1990 (Yao 1994).

Shanghai's lost decade in the 1980s mostly resulted from the basic logic of the central government at the outset of the market transition: China's economic reform must be controllable, and Shanghai, as a key industrial and economic asset, should act as a "safeguard" for national reform. That is, instead of taking a lead in drastic institutional changes, Shanghai was asked to remain relatively static at the outset of the reform in order to buffer the national economy from risks and potential shocks, stabilizing the broader process of transformation. Indeed, the initial batch of cities permitted to trial the special economic zones at the beginning of the reform, such as Shenzhen, Zhuhai, Shantou, and Xiamen, were all economically minor within the previous planned economy.

Acting as a reform stabilizer prevented Shanghai from adapting to the new market environment. For instance, under the planned economy, supplies of the raw materials of which Shanghai stood in constant need to support industrial production were comprehensively arranged by the central administration. As market mechanisms replaced bureaucratic directives, costs of sourcing raw materials soared to such high levels that many local enterprises could barely survive without increasing product prices. However, adjusting these prices was not permitted in Shanghai by the central government, which was more concerned about the risk of economic turmoil across the country at this time (see Shi et al. 1987). Second, by languishing at the back of the national reform process, Shanghai lost numerous development opportunities in the 1980s to regions that had not been highlighted in the past national economic positioning. The regions of Guangdong and Fujian, for example, got the earliest chance and took advantage of the local cheap labor to attract foreign investments in the reform and soon rose up from partaking the global labor division. Meanwhile, in Jiangsu and Zhejiang provinces, which historically contained fewer state-constructed large enterprises, township enterprises and private businesses at last began to prosper. Shanghai could follow neither of these developmental pathways in the 1980s, however.

Lastly, Shanghai lagged behind in the fiscal decentralization of the central government, a major measure in the initial stages of China's economic reform that allowed local governments to retain a proportion of tax income, stimulating more active local state involvement in economic development. In the 1980s, while most local governments began to enjoy increasing tax revenues in accordance with financial decentralization, Shanghai's historically unbalanced fiscal contribution to the nation was largely maintained. Throughout the 1980s, the fiscal burden per capita in Shanghai was 70 times that in Guangdong (Wang 2003). This stable and large fiscal turn-in from Shanghai helped to greatly smooth the course of national decentralization (Cai 1992), but held Shanghai back in development terms in the 1980s. The national decentralization that took place after the reform is further detailed in the next chapter.

The Rise of the Local Developmental State

Though Shanghai, paradoxically, had been sidelined as the stabilizer of the transforming Chinese economy, since the market reform was instigated, the local government of Shanghai never fully resigned itself to its safeguarding role, even within those lost ten years. In the face of overwhelming, centrally driven bureaucratic constraints, Shanghai officials continually requested more autonomy for the local government. After all, the performance of the local economy was not only the fundamental responsibility of the Shanghai administration, but was also closely related to the future prospects of all the major political figures involved. Calls for greater local autonomy peaked in the late 1980s:

> …Comrades Zemin, Rongji, Bangguo, and Huang Ju[1] repeatedly emphasized one issue at that time that Shanghai will not reduce one cent in the fiscal turn-in to the central government, but hopes to have local autonomous rights in the reform… As Shanghai once supported the national reform and the central leaders all acknowledged the difficulty of Shanghai by then… (the central government decided that) Shanghai should be supported, except for fiscal assistance, and thus, autonomy was finally gained. (Wang 2003: 61)

These negotiations between local and central government later resulted in a special policy guideline for Shanghai, entitled "Three Promises and Three Reforms". The three promises were aimed at fulfilling the requirements of the central authority: insisting on Shanghai's ongoing socialist orientation; following the macro-economic direction of the central government; and accomplishing fiscal commitments with surplus. The three reforms promoted local autonomous rights, which claimed that the Shanghai economy should be self-funding, self-reliant, and ahead of that of other regions.

In the 1990s, the central government began increasingly to acknowledge the need to release Shanghai to take center stage in the reform. With increasing prosperity and success apparent in the provinces that had led the way in the early days of the market transition, such as Guangdong and Fujian, the reform planners were confident that the Chinese economy had the ability to continue strengthening, and regarded Shanghai as the next potential breakthrough. In this period, Deng Xiaoping, as the main orchestrator of the transition, repeatedly underlined the urgency of developing Shanghai to spearhead the next stage of China's reform. Deng made several well-known pronouncements on this issue on different occasions, stating that "Shanghai is our trump card and a shortcut to success", while noting how "in retrospect, my largest mistake (in the reform) was not to add Shanghai in the four firstly-opened special economic zones", and "the 1990s is the very last chance for the development of Shanghai".

The new stance of the central government greatly facilitated Shanghai's realignment away from the role of economic safeguard during the next phase of the reform.

From cautious beginnings, a unique local political structure has emerged in Shanghai since the 1990s, characterized by a strong and active local government closely sponsored by the central administration. Once the Shanghai government achieved direct control over local economic affairs, its administration resembled that of many other local governments in terms of fiscal decentralization. The key difference was in how the historic prominence of the Shanghai economy had contributed to the extraordinary political power of its leaders, often winning an unusual degree of attention and support from the central administration. While the nature of the bureaucratic power structure frequently created tension between local and central levels during the lost ten years, by the 1990s central government interventions in the local economy of Shanghai no longer took the form of direct commands, but mainly arose via macro-level policies. The clearer functional division of responsibility for economic affairs also saw the Shanghai government take control over how the local economy would be managed, while central government took more of a back seat. Moreover, the relationship between the local and central governments often tended to be cooperative, as the unique economic status of Shanghai encouraged central leaders to provide special care and attention that was certainly welcomed by local officials.

In practice, local–central administrative relations in Shanghai functioned through two unique mechanisms. First, the local government actively lobbied for policy preferences from the central administration. As stipulated in the "Three Promises and Three Reforms" directive, it was unrealistic for Shanghai to seek financial assistance from the center, so the Shanghai government requested special policy treatment because of its specific needs. For instance, as early as the mid-1980s, while Shanghai was still buffered to some degree from the market transition, in order to acquire start-up capital for economic development, the local administration appealed to the central government for the right to access foreign loans—a complete impossibility elsewhere in China. Similarly, Shanghai later requested and received permission to set up a local bank, insurance company, and futures market. Most of these requests were unprecedented at the time, and unlikely to be approved had they not been submitted by Shanghai.

On the other hand, local–central cooperation also functioned in a top-down manner. Due to its perceived vast economic potential after the safeguard approach of the 1980s came to an end, Shanghai became a favored candidate for the central government to implement national development initiatives. China's new leader Jiang Zemin, who had previously been in charge of Shanghai, proposed to return Shanghai to a similar, if not even more globalized, position than it had occupied before communism. One main initiative was the Pudong economic zone, which was seen as a key project for the central government to accelerate China's market reform in the 1990s. At the 14th National Congress of the CPC in 1992, Jiang emphasized the aims of this project:

> To take the development of Pudong in Shanghai as the lead, to develop the cities alongside the Yangtze River, to quickly build Shanghai as one of the international economic, financial and trade centers, and to drive a new leap for the economy of the Yangtze River delta and the whole valley. (Jiang 1992)

Many preferential policies were subsequently offered to assist the Shanghai government in the construction of the Pudong economic zone. For instance, state-owned lands in Pudong were allowed to be leased out as a solution to fund shortages. Currency trade in RMB was also opened for foreign financial institutions in Pudong, in order to promote the local financial sector. Pudong has since successfully matured as a leading economic zone for the Yangtze delta, as the reform planners had envisaged, as well as a symbol of the revitalization of the Shanghai economy.

Local Ambition as National Breakthrough

While these wider economic and institutional changes were under way, Shanghai officials harbored long-held aspirations towards the manufacture of passenger cars, considered the most important step for the future of the local economy in the market reform. By determinedly lobbying central government, the local leaders finally persuaded Deng Xiaoping of Shanghai's potential and a trial was approved: the Shanghai government acquired a permit to run a local car project—and most importantly, to engage a foreign partner in this project. The car joint venture in Shanghai steered the Chinese automobile industry into a new age: the door previously closed to foreign capital was opened, and joint ventures soon became a legitimate strategy for technological upgrading among numerous Chinese automakers.

Shanghai's Choice to Make Cars

As early as the 1980s, Shanghai officials began to explore the possibility of local production of passenger cars. From 1981 to 1991, three consecutive mayors prioritized the car industry in their developmental agendas. The idea of launching passenger car production was first advocated by Mayor Wang Daohan (1981–1985); his successor Jiang Zemin (1985–1988) proposed that car production should be the pillar industry for Shanghai; and Zhu Rongji (1988–1991) reaffirmed the view that the car industry would be the most important sector in Shanghai. The consistent priority assigned to this sector by Shanghai's mayors was in line with the commonly held view that automobile, and specifically car, production was key to the economic revival of Shanghai in the reform. Such a selection of passenger cars as the target product by the Shanghai government was mainly based on careful observation of the emerging domestic car market in China. As detailed in the previous chapter, product diversity in the Chinese automobile industry was extremely limited in the planned-economy era, such that increasing domestic demand in the early 1980s triggered an influx of foreign cars, and import substitution was considered essential by central policymakers. For the Shanghai government, this was a golden opportunity to develop local car production.

The choice of local officials to prioritize car production also drew on Shanghai's car-making history. Prior to the reform, Shanghai had already set up one of the best-equipped passenger car assembly lines in China, though it was underdeveloped by modern standards. As shown in Table 3.2, throughout the planned-economy period, only three car models

Table 3.2 Main models of the Chinese automobile industry, 1949–1981

	Light truck	*Mid-weight truck*	*Heavy truck*	*Off-road vehicle*	*Passenger car*
1949–1956		Jiefang		Jiefang	
1957–1965	Yuejin Shanghai Huanghe	Jiaotong Hongqi	Huanghe	Huanghe	Hongqi Dongfanghong Shanghai
1966–1981	Beijing Hongwei Qiantangjiang Tianjin Shanghai	Dongfeng	Jiaotong Changzheng	Dongfeng Hongyan Yanan	

were developed in China: the Hongqi from FAW, the Dongfanghong (meaning "sun rising from the east" in Chinese, a political metaphor for Chairman Mao) from Beijing Automobile Factory, and the Shanghai from Shanghai Automobile Factory. By the early 1980s, however, the Shanghai was the only surviving model. Although FAW's Hongqi was long honored as "the national car", production was suspended in the early 1980s because of criticism of its high fuel consumption. The Dongfanghong suffered a graver fate, its production arbitrarily terminated amid the political riots of the Cultural Revolution. By comparison, production of the Shanghai model continued uninterrupted. First developed in 1958, the Shanghai reached a modest output of 50 units in 1964. From the mid-1960s to early 1980s, annual output gradually increased by over 5000 units (Committee of Shanghai Automotive Industry History 1991). Clearly, production was on a very small scale compared with modern standards, but the Shanghai was one of the most successful cars to be developed in pre-reform China. The historical achievement of the Shanghai model was strongly related to the comparatively advanced capacity for manufacturing automotive components that had already developed in Shanghai before the revolution and establishment of the new China, as part of Shanghai's historic industrial base. In the early 1950s there were about 200 auto-component makers in Shanghai, accounting for one-third of the total national output (Committee of Shanghai Automotive Industry History 1991). In addition, unlike FAW's Hongqi, which was specifically designed for high-ranking political leaders, the Shanghai was a basic model suitable for ordinary officials and able to meet a more substantial and stable demand in the planned-economy era, albeit still aimed towards official uses.

The Shanghai car-making initiative also gained support from the specific upgrading experience of the early years of reform. In 1978, the outset of the economic reform, the central government decided to trial passenger car assembly lines, and Shanghai, the only location capable of manufacturing passenger cars on a relatively large scale, was chosen for the task. Other large Chinese automakers such as FAW and Second Auto Works (SAW), later renamed Dongfeng Motors Corporation (DMC), were all busy remodeling their old truck models at the time, and thus were less interested in competing. The good fortune of being appointed to introduce a passenger car assembly line in the early years of reform may have been a historical accident, but as the central government began to promote car production on a mass scale, Shanghai was seen as the precursor for this model, and in the late 1980s, it took its place alongside FAW and SAW as one of the "Big Three".

After years of policy-induced stagnation, car manufacturing advocates began to escalate their backing of the industry as the key to reversing Shanghai's economic decline. As soon as local officials escaped the restrictions of the lost ten years, their primary focus was to renovate the outdated local industrial structure that had been put in place during the planned economy. Two pathways were proposed. The first would be to promote tertiary industry such as financial services and trade, with the goal of transforming the sector composition of Shanghai from the old pattern of secondary-tertiary-primary to tertiary-secondary-primary. Ironically, this appeared to be a retrograde step reverting to the period before the communist regime. The second pathway was to identify "pillar" sectors within the secondary industries, which could lead the reboot of the local economy. Six priority sectors were thus established in the early 1990s, with the car industry topping the list, ahead of telecommunications equipment, electric power plant equipment, petrochemicals and fine chemicals, steel, and home appliances. Both pathways were implemented in Shanghai, where car production has gone from strength to strength since Jiang Zeming and Zhu Rongji, the staunchly pro-car manufacturing former mayors of Shanghai, rose to the most powerful positions in China by the 1990s.

Origins of the Joint Venture Strategy

The decision to develop a car project was in itself no easy matter in the context of early-reform China; to figure out how to accomplish the project was a challenge of a different order. When Shanghai started its local car project in the 1980s, the Chinese car sector lagged decades behind global standards. How to catch up with the advanced technology and management systems of foreign automakers became a primary concern. Ideas of fashioning a joint venture with foreign partners gradually took hold among local officials as a necessary technological upgrading strategy.

Engaging foreign partners was a grand political adventure in 1980s China—it was contrary to the long-held communist logic of self-reliance, once a central component of the CPC's rationale of development. By criticizing the "old" China as incapable of domestically manufacturing any major industrial products, the economic nationalism gained great political legitimacy after the CPC came to power. With regard to the automobile industry, the need for an independent and complete production system was thus taken for granted: as early as October 1949, the month that the new China came into being, the embryonic government established the Ministry of Heavy Industry, among whose tasks was to initiate automobile production in China. The following year, the central government announced that China would build its own independent automobile industry, and work on the FAW facility soon followed in earnest. The pursuit of an independent, self-reliant Chinese automobile sector had been over-emphasized throughout the planned-economy era. As a result, introducing foreign investment was politically stigmatized as betrayal of the national interest, and remained so at the beginning of the market reform.

The idea of engaging a foreign partner became a material prospect in the minds of local policymakers in Shanghai largely due to the failure of the 1978 trial introduction of a foreign assembly line for passenger cars. With a view to promoting exports, central administrators heavily favored what was known as "compensation trade": the foreign client would supply the required technologies, equipment, and materials, with the Chinese partners responsible for production. These policymakers even drafted a detailed plan with specific output targets, envisaging an assembly line that would achieve annual production of 100,000 cars and 50,000 minibuses, of which 20,000 cars and 30,000 minibuses would be sold domestically and the rest exported. Such a model, if fulfilled in practice, would make perfect sense for the central government—serving the small but expanding domestic demand for cars, while simultaneously promoting import substitution and boosting foreign currency reserves through export. But the plan relied on the unrealistic expectation that foreign automakers would simply be attracted by cheap labor in China, which might be valid for labor-intensive sectors such as textiles and apparel, but could hardly function in the capital- and technology-intensive field of modern automobile production. Simply accessing a cheap labor force in China was not an attractive deal for foreign firms, whose other competitive concerns included leakage of advanced technologies, the competence of local labor and manufacturers, and potential difficulties with marketing finished products.

However, through a gradual process of negotiations between administrators of the Shanghai automobile industry and foreign automakers, a new and much more promising strategy emerged: establishing a joint venture with a foreign partner. Historical details of this procedure are well documented, especially in the narrative accounts of Jiang Tao, the main negotiator in Shanghai's first joint-venture project, Shanghai-VW, and Qiu Ke, its first director (see Wu 2009; Zheng 2007). The following extracts illustrate these intricate negotiations:

> In October of 1978, just one year after my appointment as the general director of the Shanghai Tractor and Automobile Industry Co., I was asked to lead the negotiation team for the car project. I went to Beijing for a meeting with General Motors from the United States as the first negotiator. General Motors would not accept our proposal for cooperation in compensation trade, but brought forward the idea of setting up a joint venture, in which they would provide technology, management, and capital. Under the table, I felt pleased about their suggestion, as I never considered compensation trade as a good direction for the Chinese automobile industry. Through this way, we could neither get the real technologies, nor the needed capital. But I could not argue against the State Council. Now here the suggestion just comes from the United States. I held my excitement inside, told GM to wait for further discussions, and immediately reported to the State Council. In 1978, there was no such thing as a joint venture. So many leaders hesitated to take risks... (Qiu Ke, narrative account in Zheng 2007)

> On November 9th of 1978, when Gu Ming as the associate director of the National Planning Committee asked about whether a joint venture could be applied to the car project, [Deng] Xiaoping immediately gave a positive response, 'Sure, not only for cars, the heavy trucks could also have a try'. This set the tone. The National Automobile Industry Corporation then informed Shanghai that later negotiation with foreign corporations could accept the possibility of setting up joint ventures. (Jiang Tao, narrative account in Wu 2009)

> The later negotiation was not centered around compensation trade any more, but all about joint ventures....At that time, we had negotiated with General Motors and Ford from the United States. For General Motors, we had three rounds of talks. In Japan, we met with Nissan two times. Toyota did not respond to us, and later felt regreted. For French automakers, we talked with Renault one time, Citroën about ten times, Peugeot two or three times. In Italy, we contacted Fiat. Among all of these automakers, we negotiated with Volkswagen from Germany most frequently, approximately thirty times. Finally, Volkswagen was confirmed as our partner. (Qiu Ke, narrative account in Zheng 2007)

These detailed descriptions highlight the typical workings of the local political structure of post-reform Shanghai, a local development state under sponsorship of the central government. They also reveal that the negotiating role of local industry leaders in Shanghai was key to promoting the concept of joint ventures for the development of China's car sector. These officials were well aware of the drawbacks of compensation trade and strategically steered the negotiation towards joint ventures, as was their preference. Setting up a joint venture was clearly viewed among industry insiders as a much more realistic and beneficial strategy to attract foreign automakers to China. Holding a share in a joint venture was an effective way to ease concerns among foreign investors about potential risks. More importantly, to establish a joint venture with foreign corporations was a better fit with local interests. A joint venture would improve opportunities for Shanghai to learn advanced technologies in cooperation with foreign automakers, whereas under compensation trade it was more likely that Shanghai would end up simply assembling imported automotive components from overseas suppliers: in other words, a complete-knock-down (CKD) production model.

Acceptance by Deng Xiaoping of the proposal for a joint venture was also essential. Given the zeal with which foreign capital inflows had once been condemned as a threat to national interests under the planned economy, only the highest level of power—Deng himself—could overturn this entrenched ideological position and open the way towards engaging foreign partners. In fact, joint ventures with foreign partners in the car industry matched well with Deng's overall rationale for China's economic reform. For Deng, the car industry, as well as many other sectors in China, had over-emphasized the pursuit of independence, sacrificing the chance for technological upgrading through isolation from the outside world. Deng was determined to break free of this ingrained position, prioritizing the introduction of advanced technologies and management models through policies of "opening up".

The approval of the first joint venture, Shanghai-VW, soon evolved into wider guidelines for technological upgrading in China's car industry in the 1990s, opening the door for foreign automobile manufacturers and marking a new era for the Chinese automobile industry. Other domestic carmakers seized on the success of Shanghai-VW as an opportunity to follow the same route, under the direction of the central administration (Table 3.3). In 1991, several years after Shanghai-VW, the leading centrally owned producer, FAW, set up its own joint venture with VW; DMC, China's

Table 3.3 Major joint ventures in the Chinese car industry, 1983–2004

Year	*Name*	*Foreign partner*	*Ownership structure*
1985	Shanghai -VW	VW	SAIC & others 50%, VW 50%
1985	Guangzhou PSA	PSA	Guangzhou 46%, PSA 22%, Others 42%
1991	FAW VW	VW	FAW 60%, VW 40%
1992	Dongfeng PSA	PSA	DMC 70%, PSA 30%
1993	Changan Suzuki	Suzuki	Changan 51%, Suzuki & others 49%
1997	Shanghai-GM	GM	SAIC 50%, GM 50%
1998	Guangzhou Honda	Honda	GAIG 50%, Honda 50%
2001	Changan Ford	Ford	Changan 50%, Ford 50%
2002	Tianjin FAW Toyota	Toyota	FAW 50%, Toyota 50%
2002	Beijing Hyundai	Hyundai	BAIC 50%, Hyundai 50%
2002	Dongfeng Yueda Kia	Kia	DMC & others 50%, Kia 50%
2003	Dongfeng Motor	Nissan	DMC 50%, Nissan 50%
2003	Dongfeng Honda	Honda	DMC 50%, Honda 50%
2004	Guangzhou Toyota	Toyota	GAG 50%, Toyota 50%
2004	Beijing Benz-Daimler Chrysler	DC	BAIC 50%, DC 50%

second-largest automaker at the time, selected PSA Peugeot Citroën (PSA) as partner in the following year. In 1994, the establishment of joint ventures to introduce foreign technology was officially acknowledged in China's first automobile industry policy (see State Council of China 1994).

The rise of the joint-venture model, against the backdrop of a national strategy of technological upgrading for the car sector, reflected an intriguing configuration of developmental logics. The joint-venture strategy can be viewed as a mixture of Deng's opening-up logic and Mao's economic nationalism. "Opening the Chinese car industry" was never practiced as simply introducing foreign corporations at will; for the Chinese bureaucrats in charge of the automotive industry, this introduction was the means, but not the ultimate goal. The primary aim was in fact still to build a strong and ultimately self-reliant national car industry, but one able to keep pace with global developments by learning from the substantially more advanced foreign carmakers.

With this in mind, officials designed a two-stage plan for the Chinese automobile industry (He 1994). From 1994 to the end of the twentieth century, the industry would focus on improving basic capacity and production conditions, enabling the second stage from 2000 to 2010, in which the sector should be able to catch up with advanced foreign automakers, develop an independent R&D capability, and cultivate a set of competitive domestic enterprises in the international market. More specifically, there was an official flowchart for the technological upgrading of carmakers in joint ventures (Liu 2006): in cooperation with foreign companies, domestic carmakers were expected to fully master R&D following an order for an automobile body, the production arrangement and components, chassis, and finally the engine and the complete car. In this sense, the joint-venture strategy was often referred to in the industry as "exchanging the market share for advanced technology". To further guarantee that domestic carmakers would effectively learn "advanced technology", the strategy carefully sought to avoid the risk of foreign corporations controlling the Chinese car industry via joint ventures, central policymakers insisting that no foreign corporation could hold more than half of the total share in any venture. Moreover, each foreign corporation could have no more than two domestic partners. Table 3.3 shows that these centrally imposed conditions resulted in a majority of joint ventures having a 50/50 ownership structure, with no foreign share above 50%, and no foreign automakers having more than two Chinese industry partners.

Expansion of SAIC Under Local Government

The lifting of restrictions on autonomous local management in Shanghai by the 1990s cleared the way for the Shanghai government to push for rapid development of its car industry in two key respects. First, initially based on a cluster of small local automobile factories, SAIC was organized and expanded as a corporate group owned and directed by the local government. Second, the introduction of foreign technologies through the joint-venture strategy helped the Shanghai industry to catch up with international manufacturing standards, and SAIC to finally become the largest carmaker in the domestic market. The socio-political process behind these transformations of SAIC demonstrates how the distinctive production model in Shanghai was formed from local institutional forces.

From Production Association to Corporate Group

After the market reform, and under the joint direction of the local and central governments, the organization of car production in Shanghai was transformed along a unique pathway in which the core enterprise, SAIC, was gradually assembled from members of the local production associations.

This process was in sharp contrast with the development of FAW, which, as the foundation of national automobile production, had been a leading enterprise of the planned-economy era. Once the reform started, the changes faced by FAW amounted to constant institutional re-organization in line with the evolving ideas of the central government about the relationship between state and enterprise. The pre-reform legacy of automobile production in Shanghai was totally different, however: there were no leading local automobile enterprises, but instead many small factories, most of which had emerged from earlier mechanical workshops.

Shanghai's existing production base for automobile components, dating from the 1950s, consisted of a large group of minor factories, a structure that changed little throughout the planned-economy era, with local production associations often coordinating the local manufacturers. In a typical association, a leading enterprise was appointed by the state to be in charge of hundreds of subordinate local factories. It functioned similarly to an agency of the government, managing local manufacturers to fulfill economic commands from above. This governance scheme was consistently applied in the Shanghai automobile sector: in 1955, Shanghai Combustion Engine Component Manufacturing Corporation was established as the lead institution, coordinating about 300 subordinate local automobile enterprises. In 1958, Shanghai Power Machinery Manufacture Corporation took over and held the lead position. This corporation was renamed in 1960 Shanghai Agriculture Machinery Manufacture Corporation, again in 1969 as Shanghai Tractor and Automobile Industry Co. (STAIC), a name which was retained until the late 1970s (Committee of Shanghai Automobile Industry History 1991; Committee of Shanghai Automobile Industry Record 1999). Underneath this management structure, a local production network with a specific division of labor took shape among the local automakers.

At the same time as new management models were being piloted with large enterprises such as FAW, converting production associations into modern corporate governance structures became a central theme of the Shanghai automobile sector. First and foremost, this entailed the transformation of the lead enterprise from a bureaucratic organization to an autonomous market actor. The transformation of STAIC as the lead enterprise of Shanghai's automobile production association occurred in three stages (Committee of Shanghai Automobile Industry History 1991; Committee of Shanghai Automobile Industry Record 1999). First, in the late 1970s, STAIC had been chosen as one of the 35 nationwide experimental models to trial switching from a bureaucratic institutional structure towards an economic organization under the direction of the central government. Within this new governance mode, STAIC was permitted to retain a proportion of profits for itself under a responsibility system, much like the ongoing reforms in other state enterprises. Second, in the 1980s, Shanghai Tractor and Automobile Industry Coordination Co. (STAICC) was set up, based on STAIC, after China National Automobile Industry Corporation (CNAIC) was established by the central government as the national lead corporation to integrate the automobile industry. Though the role of CNAIC as the national industry leadership body proved short-lived, as discussed in the previous chapter, STAICC was successfully established as an autonomous economic organization, and started to trial important corporate institutions, such as a responsibility system for a general manager under a board of directors. Finally, in the late 1980s, STAICC substantially restructured its internal management scheme with five divisions: automobile and motor production, tractors, components, casting and forging, and electronic equipment.

These rounds of transformation greatly altered the management structure of local automobile production in Shanghai; however an equally important outcome was that the local state seized control of the local automobile sector. In the process above, the central and local governments took on different roles in reshaping STAIC from a state administrative branch to an autonomous economic organization, with the central government initiating the reform to de-affiliate STAIC from the previous rigid system of economic planning, after which the Shanghai government took over and became the driving force. In the early 1980s, the local government strategically took advantage of the establishment of CNAIC as a chance to seize further control over local enterprises. The establishment document for STAICC issued by the local government makes it clear that the new enterprise would be under the local government:

> Under the direction of the central government, it [STAICC] is an economic organization independently practicing production, management and accounting activities with legal person status. As a bureau-class working unit, it is led by the Economic Committee of the Shanghai municipal government and also takes directions from CNAIC with regard to business operations. (Committee of Shanghai Automobile Industry History 1991: 111)

Although the 1980s had witnessed progress in the governance structure of the Shanghai automobile sector, STAICC was still a long way off from notions of a modern corporation. The manner of governance, whether between the local government and STAICC, or between STAICC and its numerous subordinates, was based on a responsibility system but still lacked a clear ownership structure. In as much as the renaming of STAICC simply added one letter, derived from the single additional term "coordination", to the previous title, the major function of STAICC remained similar to that of the previous production association, and was yet to fully mobilize its numerous subordinates. The latter were also divided between closely associated and loosely associated subordinates. With the loosely associated group, STAICC generally had no direct ownership or affiliation at all. To fulfill production requirements, STAICC often had to organize coordinated production teams among these loosely associated subordinates, rather than intervening and operating directly. Although STAICC occasionally incorporated subordinates in the 1980s, there was little effort to systematically integrate automobile production in Shanghai under one umbrella with a clear ownership boundary. In this sense, the 1980s reform only partially established a core enterprise out of a former bureaucratic institution, but left the ownership issue unresolved.

The problem was finally addressed by the local government during the 1990s, when enterprise reform in China began to place clarification of ownership structure higher up the policy agenda. In 1990, STAICC was again renamed as Shanghai Automotive Industry Corporation, close to the current name, the Shanghai Automotive Industry (Group) Corporation (SAIC). Under this new platform representing the corporatization of the industry as targeted by the central administration, the local government gradually assembled a corporate group which integrated previous subordinates through a clear ownership structure, seeking to re-establish itself as a complete corporate institution in the modern sense. The process can be understood in three major stages, as follows.

As an initial step, in the early 1990s, Shanghai Automotive Industry Corporation conformed to a new "three levels and three centers" management scheme. This scheme assigned the headquarters as the center for investment decisions, secondary leading enterprises as centers for management and profit, and subordinates as centers for production and cost control. Meanwhile, to formalize the core operations of the corporation, the local government arranged for Shanghai Trust and Investment Company to contribute stock valued at 700 million yuan, boosting the registered capital of the corporation from 2.1 billion to 2.8 billion yuan. And Shanghai Automotive Industry Corporation began to rely solely on the board of directors as its top management mechanism.

Further important changes took place in the mid-1990s, when the local government made great efforts to clarify the ownership structure between Shanghai Automotive Industry Corporation and its numerous subordinates, echoing central government's call for corporatization of state-owned enterprises. In 1994, under the new rationale of managing state-owned assets, Shanghai's State-owned Asset Supervision and Administration Commission (SASAC) authorized Shanghai Automotive Industry Corporation to manage those state-owned assets under the existing corporate structure. In 1995, the local government proposed to transform Shanghai Automotive Industry Corporation into a state-owned corporate group, under the name Shanghai Automotive Industry (Group) Corporation, abbreviated to SAIC. The local SASAC soon approved this plan and agreed that SAIC could use capital linkage to transform its subordinates into wholly owned, holding, or joint-held subsidiaries. On September 1, 1995, SAIC was officially established. Note, however, that although management responsibilities had been transferred, SAIC continued to be owned by the SASAC of Shanghai—in other words, by the Shanghai municipal government.

In the late 1990s, SAIC went public on the Chinese stock market, moving its governance and management structures into a new phase once again. The stock market in China was used by the central government as a tool to leverage capital for numerous state-owned enterprises, and the Shanghai government was keen to take advantage of the opportunity to further transform SAIC. In 1997, the Shanghai Automotive Corporation (SAC) was set up as a public corporation in which SAIC held 70% of the total share. The main asset represented by SAC was the Shanghai Automobile Gear Factory, and therefore what was floated in the stock market was actually only a component manufacturer within SAIC. Going public proved an effective mechanism for raising capital for SAIC. In ten years, the share value of SAC increased more than five-fold. In 2006, the Shanghai government started to transfer SAIC's assets related to production of complete vehicles to the public corporation. In 2011, SAIC went entirely public, with all assets securitized.

Engaging Foreign Partners

While these gradual shifts in governance and ownership were under way, joint-venture production with foreign corporations became the core strategy for technological upgrading at SAIC. China's first passenger car joint venture, Shanghai-VW, was followed by the second, with General Motors (GM), Shanghai-GM, in 1997. Shanghai-VW and Shanghai-GM made SAIC the largest car producer in China, and validated the establishment of joint ventures with foreign corporations as a viable strategy for technological upgrading. There are two aspects to the success of the joint-venture strategy. First, a major expectation that national policymakers placed on the Chinese car industry in the 1980s and 1990s was to realize import substitution, and the joint ventures in Shanghai did an excellent job of contributing a large number of domestically manufactured cars to the market. In the mid-1980s, the annual production of Shanghai-VW was about 10,000 units, but this has exceeded 200,000 units since the mid-1990s. Shanghai-GM did not reach full production until 1999, but achieved an annual output of over 200,000 units in just four years. In 2006, the two ventures achieved a combined output of about 766,000 passenger cars (see Fig. 3.3).

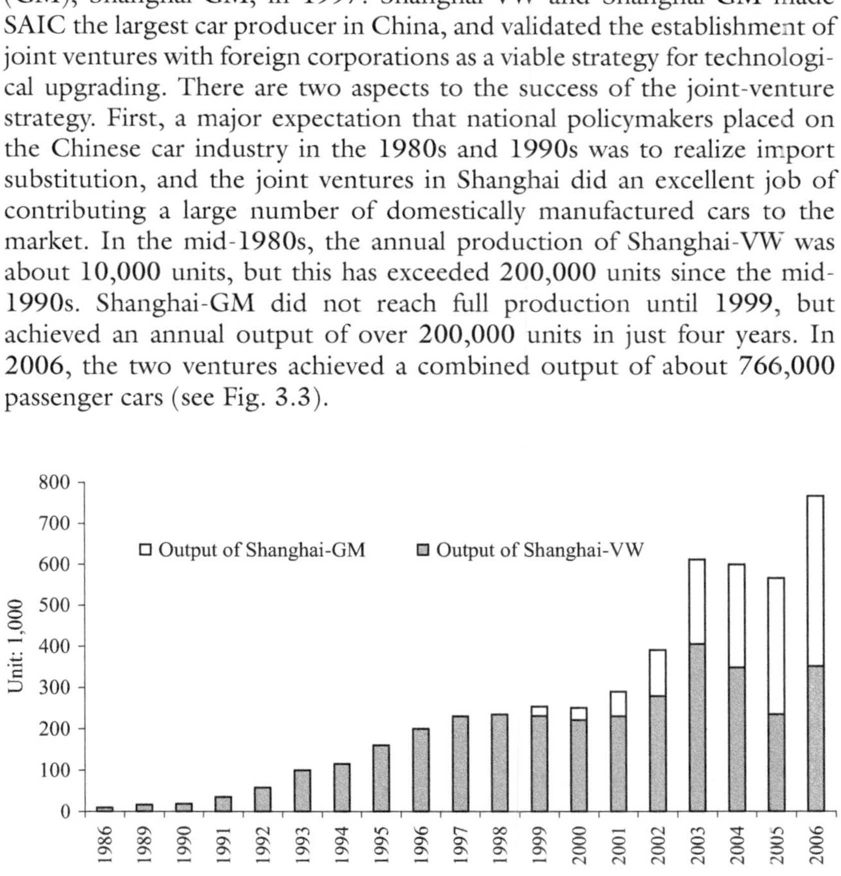

Fig. 3.3 Output of Shanghai-VW and Shanghai-GM, 1986–2006
Note: Output in 1986 is a sum from 1983 to 1986
Source: Based on SAIC Annual Reports and China Automotive Industry Yearbooks

Second, the joint-venture strategy was geared towards introducing advanced technologies from foreign automakers that the domestic industry was incapable of supplying at the time, and the two SAIC joint ventures did not disappoint the Chinese policymakers in this regard. Shanghai-VW successfully introduced a series of models from Germany, including the Santana, Passat, and Polo. From 1998 to 2004, Shanghai-GM focused on introducing the family of Buick models, commencing production of the Buick Century in 1998, soon followed by the GL8, Sail, Regal, Excelle, and Royaum. From 2005, GM further introduced its Chevrolet and Cadillac models to the Chinese market.

With the introduction of these models, Shanghai-VW and Shanghai-GM gradually delivered the most updated manufacturing technology and management system to Shanghai, narrowing the gap between the Chinese and international automobile industries. Table 3.4 shows the upgrading rate in terms of product lines at SAIC. For local production capabilities, once the ventures were properly established, the vehicles VW and GM brought to China were most often models already prepared for overseas markets. However, there is usually a time lag in the early stages. For example, as the pilot model, the VW Santana was already at the latter stages of the product cycle in Western markets when it was introduced to China, while the upgraded version, Santana 2000, was a regional model, not internationally popular. This pattern reflects a common strategy for global carmakers entering emerging markets in developing countries:

Table 3.4 Major foreign car models introduced to SAIC, 1985–2001

Year	*Model*	*Joint venture*	*Technological level*
1985	Santana	Shanghai-VW	Early 1980s technological standard; model declining in international markets when introduced in China
1995	Santana 2000	Shanghai-VW	Mid-1980s technology; model only sold in Brazil when introduced
1998	Buick Century	Shanghai-GM	Technology was comparable to late 1990s standards in North America; model was still sold in international markets
1999	Passat	Shanghai-VW	Technology was comparable to late 1990s European standard; model was still sold in international markets
2001	Polo	Shanghai-VW	Technology matched advanced global standards; Chinese model was introduced only five months after the launch of the product in Germany

Source: Based on Technology Center of Shanghai Automotive Industry (2002)

older models always tend to be favored in the transfer, as this may prolong the life of then outdated technologies in other markets, and thus indirectly save R&D costs. However, the pattern changed as the Chinese car market grew towards maturity, and competition for market share became fierce in the late 1990s, galvanizing the foreign partners to introduce leading models. A good example is the introduction of the VW Polo in 2001, which appeared on the Chinese and German markets almost simultaneously. As a result, the SAIC production technology came to be synchronized with international standards.

Though the technological upgrading enabled by joint ventures has been generally laudable in terms of the expansion of modern product lines and adaptation of advanced technology, this nevertheless comes with several important caveats. As assumed by the policymakers, the ultimate purpose of establishing joint ventures was to open the way for Chinese automakers to gradually master key technologies, and ultimately develop independent R&D capacity. In this logic, the introduction of the more advanced foreign car models was only an intermediate stage in the joint-venture strategy, and the final destination of advanced domestic manufacturing and independent R&D remained somewhat far off. Moreover, the joint-venture strategy contained hidden dilemmas, in that the highly successful introduction of foreign technology could hamper, rather than facilitate, the development of Chinese R&D capability. This was because the production of established foreign models was a shortcut to quick expansion of market share and profits for all concerned, with the potential to reduce motivation among Chinese carmakers to develop their own technologies and models—a significantly more complex and challenging task. For the foreign corporations, to promote local R&D so that the national industry could develop its own potentially competing models was self-evidently against their business interests in China. This scenario hints at the limitations that the joint-venture strategy might place on the development of national R&D, which indeed proved to be the case. Although the joint ventures were efficient at introducing technology, the Chinese side was slow to develop its own independent capabilities, and no serious attempts were made by the joint ventures to develop national models until very recently.

Though limited in terms of self-sufficient R&D, these joint ventures have greatly changed China's automobile sector. The rapid development of SAIC's joint ventures is a reflection of the sustained efforts of the local government of Shanghai at multiple levels. The joint venture with VW was achieved through tireless promotion by Shanghai officials, who played to both the business interests of VW and the policy concerns of the central government in order to reach a workable agreement. Jiang Tao, the chief Chinese negotiator for the Shanghai-VW project—later referred to as "the father of the Santana"—rose to consistently higher official positions in the Shanghai municipal government, beginning at the Shanghai First Machinery and Electronics Bureau, before being promoted to the Shanghai Planning Committee, and finally the Municipal Party Committee. The length of the negotiations—from 1978 to 1984—was mainly because of the huge gap between the expectations of central policymakers and the business concerns of VW. The central administration insisted that the project should target annual production of 150,000 units and export 80% of the total output (Ge 2007), demands that were hardly acceptable to VW, the first foreign corporation to cautiously enter the emergent reform-era Chinese car market. Local officials, though well aware of the hopelessness of the central government's directive, battled on to keep the negotiations open and prevent the collapse of a potential deal. The impasse was finally resolved when the central government was persuaded to withdraw the requests.

There were a number of other barriers in the negotiations owing to the legacy of the planned economy. For example, China's business environment in the early years of the process lacked any laws applicable to a joint venture. When the German side proposed using relevant laws from Germany, or Switzerland as a third country, local officials in Shanghai sought help from the central administration, resulting in China's first joint-venture law, which was to benefit all subsequent joint ventures. Essential guarantees required by VW were that the joint venture should have access to sufficient foreign currency quotas to import the necessary equipment, and that it should be fully supported by the central government. Local officials in Shanghai persuaded the China Bank, which had charge of the national foreign currency reserve, to allow its Shanghai branch to take a 15% share in the project. The Shanghai officials also invited the CNAIC as the lead institution for the car sector at national level to take a further 10% share, providing a strong show of support from the central administration (Jiang and Qiu 1998). These shares were later purchased back by SAIC.

A key challenge for the joint venture, led by the local government, was the arduous process of domestication of Shanghai-VW. China's two major goals for the joint-venture strategy were to achieve import substitution by increasing domestic production, and to gain technology transfer at an acceptable price. But this proved much harder than expected, as the SAIC managers found:

> When introducing the foreign model, we had taken the domestication too easily. We thought, if thousands of component makers across China were mobilized, the domestication should be realized promptly. But how could it be done in two or three days? When we really started it for the Santana, almost none of the domestic auto-part enterprises could do it without new technologies being introduced and production lines being upgraded! These thousands of auto-part producers in China were summed up as zero! This was so shocking, embarrassing, and disappointing! (Cheng 1995: 3–4)

Due to the outdated technology in China's car sector in the 1980s, most components required for the Santana model had to be imported from Germany, rather than produced by domestic auto-part manufacturers. During the first five years of Shanghai-VW, from advance production in 1983 when the first Chinese-produced Santana rolled off the assembly line until 1987, only a few simple accessories, such as tires, radios, and antennae, were manufactured in China. The domestication rate at this time was less than 3%.

In 1986, to promote the domestication of the Santana, the Shanghai municipal government established special agencies and launched numerous supporting projects. Hundreds of auto-parts enterprises across the country were enrolled. Huang Ju, then associate secretary and mayor of Shanghai, headed up a government agency to facilitate domestication. In 1988, the domestication of the Santana was highlighted by the local government as the number one project of Shanghai, and Zhu Rongji, the new mayor, began to develop a production coordination system for the Santana. By 1990, the largest project required for the domestication of the Santana, the transmission system, was successfully implemented under the direction of the local government (Committee of Shanghai Automotive Industry History 1991). As a result of these efforts, the domestic content ratio of the Santana increased from 2.7% in 1987 to 90.5% in 1996 (Fig. 3.4).

Not only did the domestication of the Santana pave the way for the future of Shanghai-VW, it also changed the entire nature of car production in China, or as one comment put it: "one project changed the whole industry". The growth of a domestic auto-parts sector and the formation of a complete auto-parts production network benefited other Chinese carmakers, including both those who followed SAIC along the path of the joint venture such as FAW and DMC, and those who were determined to produce cars independently such as Chery and Geely. The production network that had been established soon helped Shanghai-VW to increase the output of the Santana which went on to occupy the largest market share

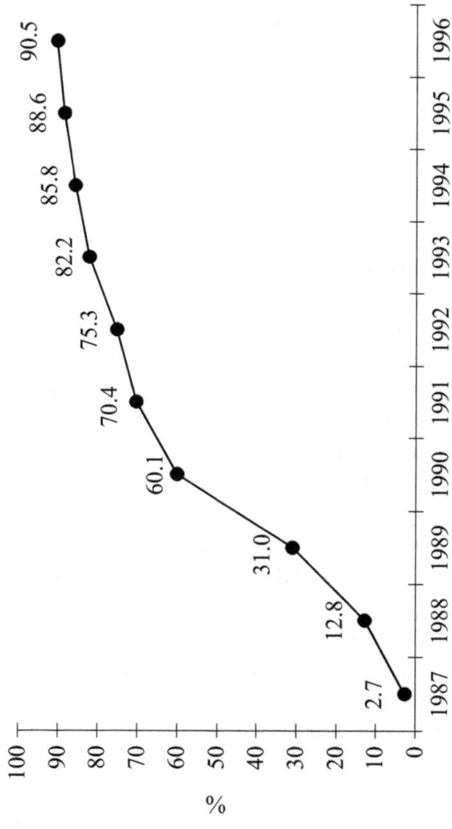

Fig. 3.4 The domestication ratio of Santana in Shanghai-VW, 1987–1996
Source: Based on Committee of Shanghai Automotive Industry History 1991, and China Automotive Industry Yearbooks

in China for the next two decades. Moreover, it greatly facilitated the introduction of other product lines through the SAIC joint ventures. For instance, the upgraded Santana 2000 introduced in 1995 immediately reached a domestication ratio of about 60% from the outset of production, while Shanghai-GM's first model, the Buick Century, achieved a domestication ratio as high as 40% in its first year of production. A more foundational achievement of the domestication project at Shanghai-VW was, perhaps, that it provided core justifications for the joint-venture strategy as a feasible pathway for the Chinese car sector. The strategy proved that when foreign models were introduced through a joint venture, Chinese automakers were able to get over primary-stage production represented by the CKD model, gradually increase domestic content, and finally achieve full domestication. This technological upgrading trajectory was exactly what joint-venture advocates had expected, though these advocates ultimately wished it to end in the birth of national cars.

The Shanghai local government was determined to pursue technological upgrading through joint ventures, even at the cost of its own established automobile production. When Shanghai-VW set up its first Santana assembly line in the mid-1980s, all the Shanghai model production lines and workshops were vacated to make way for it. In the mid-1990s, the entire production of the Shanghai was suspended when its infrastructure was required for the Santana 2000. These decisions were not easy, as the old Shanghai model was quite popular and profitable in the domestic market by this stage, peaking at an output of 8000 units in 1991, which generated a good profit (Jiang 2008). In the later 1990s, the local government became ever more ambitious, setting out a plan to build a world-class automobile plant to accommodate a new assembly line for the VW's Passat model. With a 2.2 billion yuan investment, this factory became one of the most advanced production facilities for VW vehicles in the world (Yan 2003).

The Shanghai local government was also keen to introduce new foreign partners for SAIC. The value of engaging more than one foreign carmaker was self-evident: SAIC as the weaker partner in terms of technology would have stronger bargaining power in multiple joint ventures, reducing the potential control exerted by one foreign partner. With the market position of the VW Santana well established, the prospect of finding a second joint-venture partner for SAIC began to feature on the agenda of local officials. In 1995, Huang Ju, then municipal secretary and the main director of the local car sector, formally proposed this idea to the central government and was immediately rewarded with approval, opening the door for the

Shanghai-GM agreement, signed in 1997, which was then the largest joint-venture project between China and the United States. The Shanghai government was much more experienced and effective at facilitating the progress of infrastructure construction in this, its second joint venture. In 1998, Shanghai-GM was officially designated the number one local project in Shanghai, and by the end of that year, the first Buick was driven off the production line. From the starting date of infrastructure construction, the whole procedure only took about 23 months.

Although central government was willing to compromise on some occasions, for instance eventually withdrawing its output and export demands when the initial Shanghai-VW negotiations appeared to be at an impasse, they never hesitated to issue clear instructions to defend its bottom lines when necessary. The domestication of the joint venture was considered essential by central leaders. In 1987, Zhu Rongji, then associate director of the National Economic Committee, declared that Shanghai-VW would be cancelled if its domestication ratio could not reach 40% in three years (Lu 1998). Concurrently, Yao Yilin, associate prime minister of the State Council, held a conference in Shanghai, in which he demanded that the domestication rate of the Santana model reach 25% in 1988, 50% in 1989 and 100% in 1991 (Committee of Shanghai Automotive Industry History 1991). In addition to these specific targets, the central government provided concrete policy support. For example, a tariff regulation was issued to stimulate the domestication of Shanghai-VW: in the first three years of the joint venture, the tariff for imported auto-components was set at 50%, increasing to 80% in the fourth and fifth years. If the domestication rate reached 60%, the tariff would be accordingly reduced to 60%; 48% if the ratio exceeded 60%; and 32% if the ratio exceeded 80% (Yin 1994).

Based on the steady development of SAIC after the market reform, the future of cooperation between SAIC and its foreign partners appears stable. The original contract for Shanghai-VW set the term of the joint venture at 25 years from 1985 to 2010. In 2002, a new agreement was signed in advance between the two parties, extending cooperation until 2030. For Shanghai-GM, the original contract was for 30 years, from 1997 to 2027, so this venture has at least a further decade to run. It is thus plausible that the SAIC production model featuring local state ownership and joint-venture technology strategy would be continuously practiced in Shanghai for the foreseeable future. The two joint ventures are also the solid basis of SAIC's position as the largest automobile producer in China. SAIC-VW took about 25 years to reach a cumulative output of 5 million vehicles by 2009, whereas SAIC-GM took about 14 years to hit the same level by 2011.In 2016, SAIC entered the top 50 in the Fortune Global 500 for the first time.

Case Summary

This chapter uncovers the deep institutional origins of the second production model in the Chinese automobile sector. SAIC, owned by the Shanghai municipal government, has become the largest Chinese automaker by setting up two joint ventures, with VW and GM. The chapter first investigates the unique national economic positioning of Shanghai during the planned economy, which is further demonstrated as a path-dependent force shaping the local automobile sector in the new market environment. As China's national economic center, Shanghai municipal officials carried exceptional political weight, which greatly facilitated their lobbying capabilities in the face of the central government, both prior to and after the market transition. Consequently, although in the early days of reform, Shanghai was required to hold back from the market transition that was under way elsewhere as a safeguard against the uncertainties of wider institutional changes, once these constraints were removed in the early 1990s, the local government became active in paving the way towards Shanghai's economic future. As evidenced in this chapter, proposals by local officials to use the car industry to revitalize the local economy were approved without much hesitation at central level once the policy restrictions on Shanghai's economic autonomy had been lifted. More intriguingly, the acceptance by central leaders of the locally generated, ground-breaking idea of establishing a joint venture with a foreign corporation set a new industry precedent for China, despite its political sensitivity. The resulting socio-political process around the development of SAIC was steered by a strong local government with firm support from the central administration. In this process, the Shanghai government organized the local automobile enterprise and wider production network under the umbrella of a core corporation, SAIC, as well as pioneering the first joint venture in the Chinese car industry.

Notes

1. The four persons mentioned are Jiang Zemin, Zhu Rongji, Wu Bangguo, and Huang Ju—all of whom were major municipal leaders of Shanghai, and consecutively appointed party secretary of Shanghai from 1987 to 2002. All were later recalled to central government, rising as national political leaders.

Bibliography

Cai, L. (1992). *The Choice Toward Modernization: The General Principle of Economic Development of Shanghai*. Shanghai: Shanghai Far East Press. (in Chinese).

Cao, L., Gu, G., & Liu, J. (1987). *The Study on the Ownership Structure of Production Means in Shanghai*. Shanghai: Shanghai Social Science Academy Press. (in Chinese).

Cheng, Y. (1995). Start a New Age for the Chinese Car Industry: Inspirations Form the Santana Project. In The Propaganda Center of SAIC and Shanghai Auto News (Ed.), *Rising Up by the Huangpu River: A Report from the Shanghai Automotive Industry* (pp. 1–11). Shanghai: Shanghai Wenyi Press. (in Chinese).

Committee of Shanghai Automobile Industry Record. (1999). *Shanghai Automotive Industry Record*. Shanghai: Shanghai Social Science Academy Press. (in Chinese).

Committee of Shanghai Automotive Industry History. (1991). *Shanghai Automotive Industry History: 1901 to 1990*. Shanghai: Shanghai People's Press. (in Chinese).

Committee of Shanghai Fiscal and Taxation Record. (1995). *Shanghai Fiscal and Taxation Record*. Shanghai: Shanghai Social Science Academy Press. (in Chinese).

Committee of Shanghai Local Record. (2005). *Shanghai Local Record*. Shanghai: Shanghai Social Science Academy Press/Shanghai People's Press. (in Chinese).

Ge, B. (2007). Details of Negotiations on Santana. *Auto Business Review, 7*, 118–125. (in Chinese).

He, G. (1994). Clarify Goals, Improve Strategies and Invigorate the Machinery and Automobile Industry. *Construction & Design for Project, 3*, 2–5. (in Chinese).

Jiang, Z. (1992). Speed Up the Reform, Opening, and Modernization, Achieve More for Socialism with Chinese Characteristics. *Speech at the 14th National Congress of Communist Party of China*. (in Chinese).

Jiang, Z. (2008, August 24). The Record of the Second Stage of SAIC-VW. *Shanghai Auto News*. (in Chinese).

Jiang, Z., & Qiu, K. (1998). The Reform and Opening Shed Lights on Shanghai Automobile Industry. In The Shanghai Automotive Industry (Group) Corporation (Ed.), *Flying High* (pp. 51–58). Shanghai: The Shanghai Automotive Industry (Group) Corporation. (in Chinese).

Jin, J. (1984). Changes of the Economic Position and Role of Shanghai Before and After the Liberation. *Social Sciences, 10*, 14–19. (in Chinese).

Ke, Q. (1956). Principle and Task of the Shanghai Industry. In Shanghai Planning Committee (Ed.), *Collection of Plans and Reports in Shanghai (1949–1998)* (pp. 575–582). Shanghai Planning Committee: Shanghai. (in Chinese).

Liu, Y. (2006). Working and Studying with Minister Lv. In First Auto Works & China Auto News (Eds.), *Lv Fuyuan and Chinese Automotive Industry* (pp. 54–60). Beijing: People's Press. (in Chinese).

Lu, J. (1998). My Career in Shanghai Automobile Industry. In The Shanghai Automotive Industry (Group) Corporation (Ed.), *Flying High* (pp. 36–44). Shanghai: The Shanghai Automotive Industry (Group) Corporation. (in Chinese).

Ma, B. (1988). The Historical Ideas and Practices of Shanghai Economic Development. *Shanghai Economic Studies, 5*, 74–79. (in Chinese).

Planning Committee of Shanghai. (1957). *The Report of Shanghai Second Five Year Plan*. Planning Committee of Shanghai. (in Chinese).

Shanghai Zhengda Institute. (2002). *The Strategic Position of Shanghai After WTO*. Beijing: People's Press. (in Chinese).

Shi, J., Xie, B., Cheng, H., & Jiang, W. (1987). The "Shanghai Illness" in the 1986 Economy and Its Treatment. *Shanghai Economic Review, 1*, 60–61. (in Chinese).

State Council of China. (1994). *The Industry Policy of the Automobile Industry*. State Council of China. (in Chinese).

Tang, L. (2001). The Changes of Strategic Ideas of the Central Committee of CPC on Shanghai after the Establishment of New China. *Party and Government Forum, 6*, 18–20. (in Chinese).

Technology Center of Shanghai Automotive Industry. (2002). *The Discussions on Improving the Capacity of Self-Reliant Innovation in SAIC Group*. Shanghai: Technology Center of Shanghai Automotive Industry. (in Chinese).

Wang, Z. (2003). The Evolution and Prospect of the Economic Development Strategy of Shanghai. In A. Wang (Ed.), *The Development of Shanghai and the Capacity of Governance* (pp. 57–77). Shanghai: Wenhui Press. (in Chinese).

Wu, Q. (2009). *Unusual Early History of the Chinese Automobile Industry Memorized by the Father of Santana*. http://auto.gasgoo.com/News/2009/01/07101532153249138454.shtml. Accessed 10 June 2017. (in Chinese).

Yan, G. (2003). *The Santana Model Witnessed the Growth of the Chinese Automobile Industry*. http://auto.sina.com.cn/news/2003-07-28/42011.shtml. Accessed 10 June 2017. (in Chinese).

Yao, X. (1994). *Fifteen Years of Shanghai in Reform*. Shanghai: Shanghai Social Science Academy Press. (in Chinese).

Yin, T. (1994). Policy of Domestic Content Requirement and Import Regulations. In Ministry of Machinery Industry of China (Ed.), *Material Collection of PSE'94* (pp. 41–45). Beijing: Ministry of Machinery Industry of China. (in Chinese).

Zheng, Z. (2007). *Li Shufu, Automobile "Maniac": Decoding the First Private Carmaker*. Beijing: China Citic Press. (in Chinese).

Zhu, J., Gan, H., et al. (1999). *Shanghai General History (Volume 12): Modern Economy*. Shanghai: Shanghai People's Press. (in Chinese).

CHAPTER 4

Chery Auto, Champion of National Brands

In the late 1990s, Chery Auto emerged from a lengthy period of central neglect and, at times, outright opposition to become the unlikely champion producer of national cars in the Chinese automobile market. Starting as a small firm in Wuhu city, Anhui Province, with the improbable dream of producing market-leading cars, but with zero practical experience, Chery developed in a totally independent manner, with no foreign partner and minimal sectoral input compared with the established producers. Behind this miraculous success story lies the story of a project to develop national-branded passenger cars, established in secret to evade central regulations, and operated by the local governments of Wuhu city and Anhui Province. With a strategy of independent technological upgrading under local government management and ownership, Chery Auto represents the third production model—a development which has not been analyzed in previous literature on the Chinese automobile sector.

The chapter opens with an analysis of the economic positioning of Anhui Province under communist central planning as a raw material-producing region serving the requirements of industrialization in other areas. The effects of fiscal decentralization in the 1980s, which loosened the reins of hierarchical control to allow for the advance of active local developmental states, are then discussed. The previous economic positioning of Anhui, not a national focus of industrialization, made it possible for this region to enjoy significant autonomy after the market transition. A unique local political arrangement, encapsulated in the idiom "businessman with red hat", is

Q. Feng, *Variety of Development*,
https://doi.org/10.1007/978-981-10-5912-4_4

further explored in this section. This term refers to the direct involvement of local officials in the emerging market, as business managers with a single motivating purpose: to effectively mobilize resources and expedite the growth of the local economy. The former planned-economy role of Anhui exporting raw materials is then related to the developmental ambitions of local bureaucrats, who were determined to get rid of the old national positioning. To produce cars became the local dream after the market reform. These ambitions proved a recipe for disharmony with the rationales of the central authority in strictly regulating the Chinese automobile sector. The final section details the socio-political process of conflict and compromise that occurred between the local and central governments surrounding Chery Auto, which eventually emerged as a successful local state-owned enterprise, producing national-brand cars with self-sufficient technology.

Anhui, a Resource-Contributing Province

In the era of planned economy, when the central government assigned specific economic duties to different geographical regions of China to serve its national industrial plans, Anhui Province, like other local state apparatus, functioned under strict central command. While China's northeast was assigned the role of national heavy industry base, and Shanghai was considered the center of the national economy, Anhui was given the ancillary role of supplying raw materials to support national industrialization, mainly exporting agricultural products and minerals to other parts of China. Consequently, local industrialization received less attention from central planners.

This economic position on China's economic map derived from Anhui's geographical and historical attributes. An inland province located in southeast China, Anhui is known for its abundance of natural resources, with deposits of pyrite, calcite, and alunite ranking among the top three provinces in China, while other minerals such as coal, iron, and copper rank among the top ten. Plentiful mineral reserves did not nurture a strong manufacturing sector in Anhui, however; on the contrary, the local economy had been dominated in the recent past by agriculture. When the communist regime was established in 1949, Anhui's agricultural output value was almost four times its industrial output value (Economic and Trade Commission of Anhui Province 1999). A province intersected by the Yangtze and Huai rivers provided easy and cheap navigation routes through which Anhui would contribute grain, coal, and many other resources to nearby more advanced manufacturing regions, such as Shanghai. In this sense, proximity to Shanghai was less a blessing for Anhui, than a curse.

Under the pre-reform centralized mechanism of financial resource allocation, the investment Anhui received from central government was mostly channeled into agricultural or mining-related projects. One of the most representative and lengthy nationally funded projects in the province was the renovation of the Huai river. Commencing in 1950, the Huai river project was the first attempt by the communist government to systematically renovate a major river; investment has continued to the present day, aiming to combat flooding and boost local agriculture. Minimal support was garnered from above to fuel local manufacturing, and the majority of local industrial projects were linked to mining. In the first five-year plan (1953–1957), several large mining and metals enterprises were established in the province, including Maanshan Iron Factory, Tongguanshan Copper Mine, and Huainan Coal Mine, while the second five-year plan (1958–1962) added Huaibei Coal Mine, Chaohu Cement Factory, Hefei Chemical Factory, and Huainan Fertilizer Factory. These enterprises formed the skeleton of Anhui's industry in the planned-economy period, typified by extraction and refinement of resources. The investments came with a rigid principle of economic command attached: that primary products made in Anhui should be continuously shipped out to support other regions. "Every year, Anhui would export more than one billion *jins* of grain,[1] five million tons of coal, several hundred thousand tons of pig-iron and steel, several tens of thousands of tons of blister copper, and large amount of electricity to support the economic development of Eastern China and the whole country" (Su and Ou 1987: 36–7).

Agriculture thus became the hallmark of the province, and it was not until the 1980s that the local secondary sector began to outweigh the primary sector. Anhui was, meanwhile, locked into a problematic trajectory of resource-based industrialization. In comparison with manufacturing, the extractive sector, despite being a branch of heavy industry, entailed less technology, lower value added, and weaker capacity to accumulate financial capital. This was especially true since prices of the resulting primary products were tightly controlled by central government. Anhui therefore harbored almost no advantage with which to compete for major manufacturing projects funded by central government and was, unsurprisingly, ignored in the designs of central administrators considering areas in which to concentrate economic development. Prior to the 1970s, when the central government decided to establish major industrial projects in the hinterland, Anhui was categorized as a coastal province; after the 1970s, when the focus of national planning shifted back to the coastal areas, Anhui was redefined as an inland province (Su and Ou 1987). As a result, the province's fiscal balance sheet progressively deteriorated. Fiscal input from central government was only 2.5% of the national total in 1984, although 5% of China's population resided in Anhui (Xu 1986). Central investment in the province was also very low in comparison with elsewhere. In 1978, fiscal expenditure per capita in Anhui was only 38.6 yuan, the lowest rate in China, while Shanghai enjoyed the highest rate of 236.8 yuan per capita (Wu 2008). Anhui thus became one of poorest provinces in China before the market reform, with per-capita GDP ranking in the lowest quarter of Chinese provinces in 1978.

Although the planned economy was not always a completely rigid top-down system, the will of the central authority prevailed over that of local governments. A typical example in Anhui was the trial of an agricultural responsibility system during the planned-economy era. In the early 1960s, aiming to restore local agricultural production that had been all but destroyed by the People's Communes,[2] the provincial government of Anhui experimented with a responsibility system that allowed households to farm autonomously beyond collective arrangements. The new method was initially acknowledged by the central authorities and quickly spread across the province: by the fall of 1961, about four-fifths of all production units in Anhui had adopted the new policy (Liu and Zhou 1983). Applauded by local peasant farmers, the responsibility system significantly increased agriculture output, but was overturned just a few months later, when national planners changed their minds. The responsibility system was criticized as a political mistake, and abolished in 1962. Zeng Xisheng, the provincial leader and an ardent advocate of the responsibility system, was removed from his position. The withdrawal of the system had disastrous consequences for productivity. From 1957 to 1978, grain output per person increased negligibly in Anhui, from 1542 *jins* to 1812 *jins*, which was lower than the productivity recorded in the Han Dynasty two thousand years previously (Wang 1989). Local initiatives were thus suppressed, even when they might represent a step in the right direction. It was not until 1978 and the beginning of market reform that the agricultural responsibility system was restored in Anhui.

Decentralization and "Businessman with Red Hat"

A key decision enacted by the central authorities at the outset of the market reform was to retreat from local economic affairs. Anhui was not in any way a priority region in the national developmental agenda when decentralization started. Its low status on the national economic map contributed to a relatively loose central–local relationship in Anhui, paving the way for local governments to take the lead in their own economic development. This local political structure brought about a unique approach by local officials to the promotion of economic growth, and the term "businessman with red hat" was coined for an individual who assumed the dual roles of government official and business manager. It was the newly autonomous and proactive local governments in Anhui Province that finally gave birth to Chery Auto.

Decentralization, Pre- and Post-reform

If decentralization is understood as the transfer of administrative powers from central to local levels, then decentralization in China's market reform was nothing particularly new, but a familiar routine that had already been deployed in several phases of the planned economy. Prior to the reform, central government had constantly used decentralization to adjust the central–local relationship. At the point of initial establishment of the planned-economy system, the voices of central administrators were paramount. Economic administration was highly concentrated at the central level, and local economic affairs such as planning, investment, construction, management, and taxation were all directed from above. During the first five-year plan, the fiscal income of the central government was about 80% of the national total; at the same time, its fiscal expenses accounted for about 75% of the national total (Jing 2008).

However, this initially unyielding top-down governance generated complaints, as local economic initiatives were inevitably quashed and local officials saw themselves as puppets operated by different branches of central government (An and Ren 2008). Haunted by this problem, national leaders constantly sought to adjust the governance system by providing a certain degree of autonomy to local levels. Two major waves of decentralization occurred during the planned-economy period. The first took place in the late 1950s alongside the instigation of the Great

Leap Forward, when Chairman Mao proposed that local governments ought to have more freedom in economic planning, enterprise management, and taxation. The second wave was launched to accompany the Cultural Revolution (1966–1976), when Mao again appealed for the central government to recede from intervening in local affairs. The impact of these early rounds of decentralization should not be overstated, however. Within the planned-economy system, the downward transfer of administrative powers was more of a re-delegation of certain duties, and the local autonomy gained in the process remained bounded by delineated directions from central government. Taking Anhui as an example, during the first wave of decentralization, local governments largely invested in projects related to the Five Smalls—five categories of small-scale industrial project relating to steel, mining, cement, machinery, and fertilizer, which were already highlighted by central planners as the core goal of the national Great Leap Forward movement.

Decentralization after the market reform, with power being redistributed in an environment of market transition, was quite different from the two previous waves, and involved changes in fiscal and taxation institutions (Fig. 4.1). In the 1980s, under a scheme called "eating in different pots", channels of fiscal spending and taxation were divided into two domains, comprising enterprises charged separately by central and local governments, with fixed transfers between the two groups of enterprises. In the 1990s, a new tax-sharing system was further developed, in which the central and local governments applied different categories of taxation. Some taxes, such as consumption tax, came under the remit of the central authorities, while others, such as business tax, were a matter for the local authorities. A third category, which included value-added tax, was shared between the two institutional levels.

The fiscal decentralization of the reform era unleashed local governments as economic actors in the emerging market. Freed to a great extent from central command, with independent fiscal resources, local governments became largely autonomous in their pursuit of economic growth, and thus responsible for any market gain or loss. Officials with good economic performance received political credit, creating strong incentives for local governments to participate in market activities. Thus, as the more extensive post-reform decentralization kicked into gear, officials of local government departments became highly motivated in the local economy, providing public goods, issuing industrial policies, and even directly establishing businesses. Such a role on the part of the local governments was

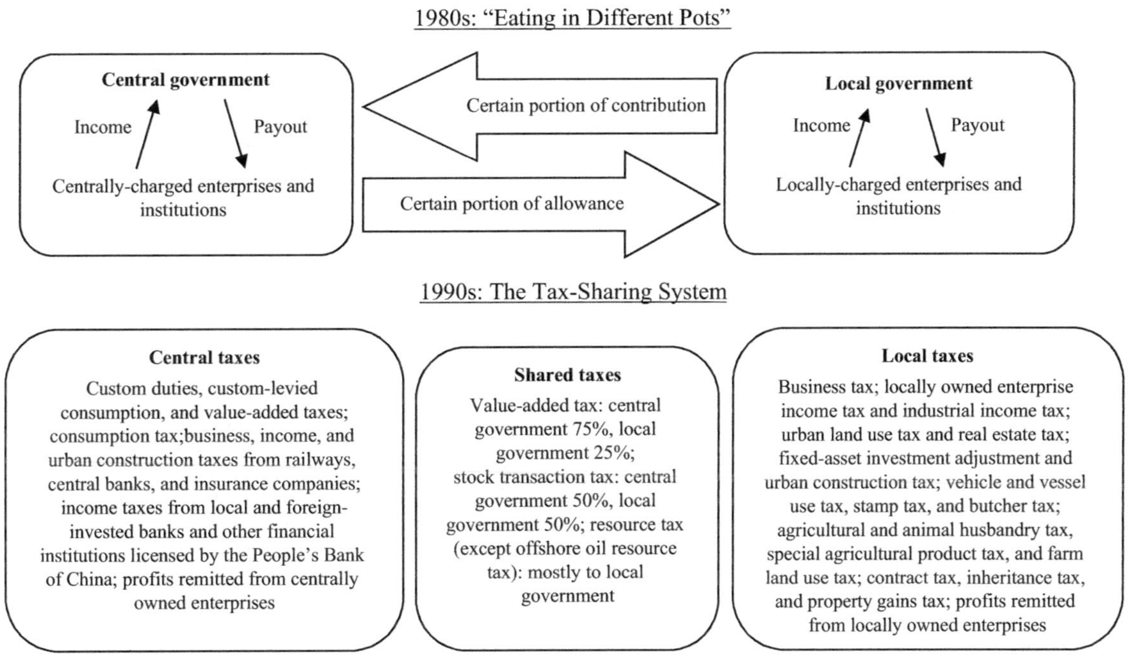

Fig. 4.1 Fiscal decentralization of China in the 1980s and 1990s
Source: Based on Zhao and Zhang (1999) and Jing (2008)

conceptualized as "local corporatism", and considered an institutional solution to weak markets in the initial stages of the transition economy (Oi 1992; Nee 1992). In practice, active local governments constituted a major growth engine for China's economic development as the market reform gained traction.

The Rise of the "Businessman with Red Hat"

The active involvement of local governments in the market led to an important development: the political and economic roles of state institutions often became merged. While elsewhere this may only have extended to local government agencies intervening in markets in certain ways, in Anhui, the combination of political and economic roles became so pervasive that it took on a form of its own, termed the "businessman with red hat". "Red hat" literally refers to a government official or political position, and thus a "businessman with red hat" refers to one person simultaneously holding both government and business positions. The ranks of the so-called "businessman with red hat" swelled dramatically across the cities and counties of Anhui Province following the market reform. Table 4.1 lists some key examples, showing that many high-level officials were appointed chief executives in local enterprises, without having to abandon their political affiliations; and similarly, successful businessmen were occasionally recruited by the government to local political leadership positions, with no requirement to drop their business affiliations.

The hometown of Chery Auto, Wuhu city, provides a particularly good demonstration of this phenomenon, including Zhan Xialai, often named the "Father of Chery Auto", who was also among Wuhu's top political leaders (Table 4.1). The city featured an institutionalized appointment system, in which a large number of local officials occupied business positions under government arrangements. The state-owned Wuhu Construction and Investment Corporation (WCIC), set up in the late 1990s, was a key platform through which local officials were able to reinvent themselves as businessmen. For the local government, WCIC offered not only a channel to gain access to market opportunities, but also a training program for budding business managers:

> The firm is located within the building of the Finance Bureau of Wuhu. The previous senior vice-mayor is the chief executive … most of the board of directors are municipal leaders. And the general and vice managers are directors and vice-directors from the Finance Bureau and the Land & Resource

Table 4.1 "Businessmen with Red Hat" in Anhui Province

City/county	*Name*	*Government position*	*Sequence*	*Corporate position*
Wuhu	Zhan Xialai	Secretary of the municipal government	⇨	Chief executive of Chery Auto
Woyang	Liu Junqing	Associate secretary of the county committee	⇦	Chief executive of Shuanglun Group
Woyang	Ma Jinhua	Associate dean of the county People's Congress	⇦	Secretary of the Party Committee of Shuanglun Group
Jieshou	Zhao Chaobin	Chair of the municipal Political Consultative Conference	⇦	Chief executive of Tian'an Group
Jieshou	Wang Xianyi	Mayor	⇨	Director of the Shahe Winery
Jieshou	Pei Jie	Associate mayor in charge of industry	⇨	Chief executive of Biquan Group
Linquan	Liu Sumin	Member of the Standing Committee of the county	⇦	Director of Liquan County Fertilizer Factory
Bozhou	Wang Xiaojin	Member of the Standing Committee of the county	⇦	Chief executive of Gujing Group
Qianshan	Yang Xuxi	Associate head of the county	⇦	Chief executive of Quanli Group

Source: Based on Fan (2003)

Note: Direction of arrows in table indicates temporal sequences for economic versus political positions

> Bureau … Organization Department of the Wuhu municipal government considered these 'businessmen with red hats' enrolled in WCIC to have field-work practice. Every two years, new officials would be sent out to WCIC and old ones would be called back. Educating a large number of 'businessmen with red hats', WCIC is thus called 'the cadre school of Wuhu'. (Zhu 2004)

In post-reform China, most local governments quickly became aware of their own economic interests and performed actively in the market, so a lingering question is: why did Anhui and Wuhu adopt such a radical approach, with the direct involvement of local officials in business management? The answer lies partly in the fact that the unique economic position of Anhui in the communist era made it impossible for other players to compete with local state actors when the market reform started. The retreat of central government intervention in local economic activities in Anhui during the reform was much more complete compared with the Northeast, where central authorities had retained strong attachments to key enterprises such as First Auto Works (FAW). A large power vacuum could therefore be freely colonized by the Anhui authorities. Meanwhile, unlike the strategic economic position enjoyed by Shanghai, local government departments in Anhui had much lower expectations of receiving extra sponsorship from above, and were compelled to take matters completely into their own hands. A further factor was the resource- and agriculture-dominated local economy, which confined most of the Anhuinese labor force to farming or mines, placing local private business in a weak and underdeveloped position in the early days of the reform. Outflows of labor from the agriculture and extractive sectors in Anhui were slow, and the private economy was stagnant. This was in clear contrast with nearby Zhejiang province, where the underdevelopment of local agriculture facilitated an active private economy after the reform, and helped nurture Geely Auto (see Chapter 5). Finally, as an inland province, Anhui was not particularly attractive to foreign investors, which gave little room for transnational corporations to take the lead in developing local markets, as had been the case in coastal provinces such as Guangdong. Even in the late 1990s, Anhui's total foreign capital in actual use accounted for only about 1% of the national total. All these factors in combination made local governments the only candidates to drive the expansion of local markets in Anhui, especially for ambitious industrial projects such as Chery Auto.

Probably of greater importance than all these factors was that the concept of "businessmen with red hat" was deliberately used as a vital developmental strategy by local officials. In the perception of local political leaders, after long being sacrificed for the greater good of industrialization in other regions such as Shanghai, Anhui was already at a comprehensive disadvantage at the beginning of the market reform, and the dual role of the local political and economic elites was a pragmatic and strategic compensatory measure to catch up with other regions. This was perceived as advantageous in propelling the local economy forwards, as commented on by an anonymous local official:

> In the stage of accumulating initial capital, the combination of the roles of officials and businessmen, and certain measures of the planned economy, could indeed have great efficiency and effectiveness. As long as it is benefiting socioeconomic development, we should accept special measures in special periods, adapting to local conditions. There is no need to be arbitrary here. However, we admit, sooner or later, such an arrangement will be cancelled and replaced by a more normal scheme. (Xu 2004: 54)

The similar logic of Wuhu city officials with regard to institutions is exemplified by their attitude to WCIC. This corporation was originally set up to receive a large loan from the National Development Bank in the late 1990s. However, it soon became a major mechanism for local elites to practice dual political and economic roles.

> (To set up WCIC is) to catch a policy opportunity, to increase the state's capacity of coordinating management over construction funds, to prevent overlapping interventions of different state branches, to avoid repeated investments and waste of resources, and thus to fully take advantage of the loan. (Xu 2004: 54)

Officials further acknowledged that this arrangement was akin to a monopoly, with all the advantages held by the dominant political-economic elite. However, they never hesitated to publicly highlight the necessity of such a monopoly to effectively pool and allocate limited local resources to major projects of public or government interest. Here is a typical comment on WCIC by Wang Bin, both associate manager of WCIC and vice-director of the Finance Bureau in Wuhu city:

> Assured by reputation, capital, and land of the government, this corporation has raised funds from multiple channels, and has acquired tremendous socioeconomic returns through a monopoly of city resources, insistence on market orientation, and operation of financial capital. The WCIC openly raises, manages, and accumulates capital for the government. And the gained profits are used for key projects of infrastructure construction in our city.... Such monopoly generates effectiveness! Such monopoly generates efficiency! (Zhu 2004)

This interchangeability between the administrative and economic roles of government employees was, rather unsurprisingly, politically sensitive, despite the supposed gains in effectiveness and efficiency. First of all, in an emerging market, the dual roles of government officials with business affiliations was considered a breeding ground for corruption, as such officials could enjoy advantages as relative pioneers of opportunities for rent-seeking. In the eyes of the public, therefore, WCIC represented a high risk of corruption on a mass scale. In addition, to allow local governments to directly compete with private corporations for profit was also considered a violation of free market principles. This was indeed unfair to normal corporations that lacked "red hat" official directors. In reality, however, these criticisms seem to have been unfounded in Wuhu city, where evidence of corruption surrounding "businessmen with red hat" proved rare on investigation. Meanwhile, the economic achievements under this questionable strategy were undeniably impressive. As will be discussed in the following sections, the achievements of "businessmen with red hat" in Wuhu city are perhaps best exemplified by the Chery Auto project, in which proactive local officials paved the way for a car plant to rise effectively from thin air.

Despite this remarkable accomplishment, public pressure would eventually see to the abolition of the "businessman with red hat" in Anhui. In 2004, Zhan Xialai and other political leaders in Wuhu resigned in succession from their business positions, despite no claims of corruption ever being laid at their doors. By then, Chery Auto's leading position in producing national cars in the domestic market had been established beyond question. Importantly, the political influence of these officials over the local economy persisted, though in a different form. Enterprises previously directly managed by officials remained in local government ownership, albeit no longer overseen by the same individuals.

Local Dreams Versus National Regulations

In the early days of the market reform, Anhui Province was locked into a skewed economic structure that had formed during the planned economy. Local officials labored to break the entrenched shackles of the planned economy and forge a new developmental trajectory. After multiple rounds of debate and discussion among local political leaders, car production was finally accepted as a necessary step to catalyze local industry. Such a plan was at the time considered a serious mistake in the eyes of the central administration, however. Since the 1980s, national industrial policy for the automobile sector had strictly regulated new market entrants in order to promote industrial concentration. The embryonic dream of car manufacturing in Anhui Province evidently offended the national will. This basic clash in the configuration of local development ideas underpinned the tortuous process of Chery Auto's emergence in the late 1990s.

Anhui's Developmental Trap

If market reform was a trigger for change in Anhui, it was not in itself the change: post-reform Anhui remained firmly fixed to its existing trajectory of economic development throughout the 1980s and 1990s. Even in the late 1990s, such were the deeply entrenched path dependencies that the structure and characteristics of the Anhui economy appeared not to have substantially altered. By the close of that decade, Anhui recorded annual output of 25 million tons of grain, 1.5 million tons of oil plants, and 250 thousand tons of cotton, ranking nationally fifth, third, and sixth, respectively (Zhou 1998), whereas industrial output value per capita was only a quarter that of neighboring Jiangsu and half that of Zhejiang, the two adjacent provinces (Social Science Institute of Anhui Province 1999).

In seeking to forge a new path, Anhui policymakers encountered developmental bottlenecks throughout the 1980s and 1990s. Long-run prospects for Anhui appeared dim: its position in national economic rankings was static, as a moderate region in terms of total economic output. Among the 28 provinces and three provincial-level cities (Beijing, Shanghai, and Tianjin), from 1985 to 1999, Anhui's GDP floated at around 13th place, while provincial GDP per capita was around 20th (Table 4.2). Behind the scenes the economic structure of Anhui continued to be dominated by agriculture. In this period, Anhui's agricultural output remained in the national top ten; at the same time, the rankings of secondary and tertiary industries were unexceptional compared with the growth of these sectors in other provinces (Table 4.2). To make matters worse, like any agriculture-based economy, Anhui was highly susceptible to unfavorable weather changes, and often underwent major economic shifts due to climate variability in the 1980s and 1990s (Hu et al. 2001).

It was hard enough for the resource-oriented local economy to exit this historical loop, but harder still to remove the stereotype of agriculture from the dominant mindset of local policymakers, a major factor in the continuing low economic status of the province. Behind the sectoral lock-in that held the Anhui economy in stasis were stubborn local

Table 4.2 Economic ranking of Anhui Province in China, 1985–1999

	1985	*1990*	*1995*	*1999*
Total GDP	13	13	14	13
GDP per capita	20	23	20	21
GDP of primary industry	8	8	9	8
GDP of secondary industry	15	13	12	14
GDP of tertiary industry	14	15	16	15

Source: Editorial Committee of Anhui Yearbook (2000)

Note: Ranking was based on 28 provinces plus the Beijing, Shanghai, and Tianjin municipalities officials lacking industrial ambition, whose preference for agricultural traditions would not fade easily. For them, agriculture had become a local comparative advantage and thus should always be prioritized in the development agenda, even after the market reform. Throughout the 1980s and 1990s, these policymakers doggedly clung to agricultural bias in their economic planning, and policy documents were thick with statements about how important it was for the local economy to concentrate on farming, exemplified in these statements from the series of Anhui provincial yearbooks:

> "(The goal) is to develop Anhui province as the national base for commodity grains and subsidiary agricultural products" (1984); "Different levels of government should all put agriculture at the top of their economic activity" (1990); "Our province is an agricultural province and agriculture is our traditional advantageous industry, therefore, at all times, we should take agricultural and rural work as priority" (1994); "(We should) strengthen the fundamental role of agriculture, guarantee the stable development of the rural economy, and always insist on agriculture as the priority of the local economy" (1997).

However, over the same period, dissenting voices grew progressively louder. Worried about the sectoral bias, some local policymakers and scholars proposed that a shift to industrialization was unavoidable for the future of Anhui. In their words, while Anhui had long been an "agriculture-industry" province, its next developmental stage should reverse toward an "industry-agriculture" province (Gao 1986; Su and Ou 1987). Thanks to these advocates, the developmental logic of Anhui gradually adjusted in the 1990s. Local economic planners slowly but surely started to recognize the necessity of promoting industrialization: the eighth five-year plan of Anhui (1991–1995) announced a clear intention on the part of the state to build up an industry-dominant economic structure; the ninth five-year plan (1996–2000) further stressed the need to reinforce the role of industry in the local economy, and to realize primary industrialization in the new century. Against initially long odds, efforts to renovate the antiquated economic structure of Anhui were proving successful (Fig. 4.2). By the end of 1990s, secondary industry accounted for almost half of the total provincial economic output, whereas the proportion held by agriculture declined to about one-quarter.

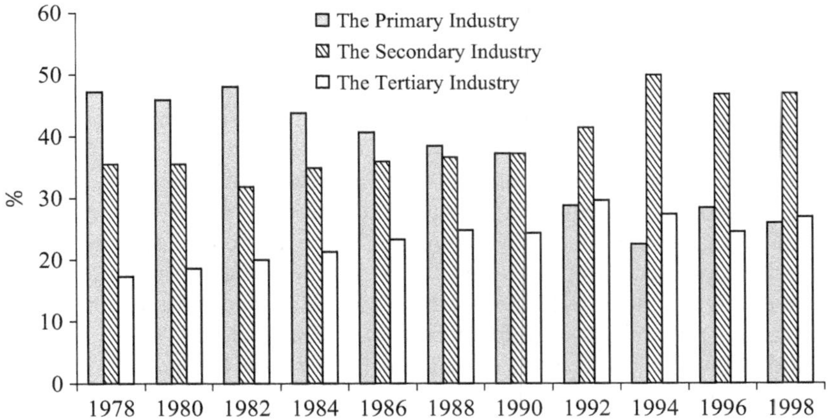

Fig. 4.2 Industrial structure of Anhui, from 1978 to 1998
Source: Based on Anhui Yearbooks

Within policy ideas to revitalize industry ahead of agriculture, "internal adjustment of industrial structure" became a central theme. Underdevelopment of industry in Anhui was largely attributed by some local policymakers to the slow adjustment of the old industrial structure. Accordingly, local officials reprioritized focal sectors for modification, based on the straightforward criteria of hitting higher value added, higher levels of technology, and higher profits. Machinery and electronics were the favored bets, for which the internal adjustment of industrial structure made significant progress. Within heavy industry, the ratio of mining, raw materials, and manufacturing shifted from 13.3: 47.6: 39.1 in 1981 to 7.4: 37.5: 55.1 in 1998; while in light industry, the proportion of output accounted for by processed agricultural products declined from 80.2% in 1981 to 66.9% in 1998 (Ge and Xia 1999). Aware of its limited fiscal resources, the local government planned to "release the leading birds first", investing in a limited number of focal, carefully chosen projects. As the "leading birds", machinery was afforded full official support until the mid-1990s, including manufacture of lift trucks, numerical-control machine tools, forging machines, electric motors for sewing machines, and motorcycle odometers (Bao 1992; Zhao 1995). Finally, in the latter half of the 1990s, local policymakers began to see the car industry as a potential "leading bird".

Cars, a New Solution for the New Anhui

For some outsiders, the prospect of setting up a car plant in Anhui was either highly ambitious or plain irrational, as no serious attempts at car production had ever been made in this region. But for local officials, this initiative was not the stuff of fantasy. While quite a shift in emphasis from the previous focal sectors, automobile manufacturing was not a totally new field for the province, as Anhui had developed small-scale truck production during the Cultural Revolution, when the central administration urged all local economies to develop a complete industrial system. The legacy of truck making, limited though it was, proved quite promising in the initial stage of market reform: in the 1980s, small and midsize trucks branded Jianghuai and made in Anhui achieved relatively good sales.

Trucks were not what local policymakers had in mind, however—they were already preoccupied by the significant potential for a domestic car market. With Chinese households gaining new purchasing power on the back of rapid nationwide economic growth and development in the 1990s, market analysts agreed that the industry was on the cusp of a massive demand for passenger cars (National Planning Committee 1997). The proposal to produce cars locally seemed to emerge almost simultaneously from two levels of government, the Anhui provincial government and the Wuhu municipal government, and the Chery Auto project was established with the mutual acknowledgment and support of both levels.

Provincial officials in Anhui heralded car manufacturing as a new answer not only to the establishment of a local automobile sector, but to the future of the entire economy of Anhui. In 1996, when drafting the ninth five-year plan of Anhui Province, vice governor Wang Yang (1993–1999) made perhaps the first official appeal for developing a local car industry. The proposal openly confirmed the ambitions of Anhui officials with a set of well thought-out justifications. For Wang Yang (1996a), the car industry was a good choice for Anhui, because (1) passenger cars had demonstrably been chosen as a pillar industry by many developed countries to boost economic growth; (2) carmaking has unique merits such as long and extensive production chains, technological intensity, high value added, and large market demand; and (3) the car industry could push Anhui onto a fast and sustainable track of development, fundamentally changing the structural bias in Anhui's economy. Wang's last point linked the development of car manufacturing to the key direction of Anhui economic policy in the late 1990s: "internal adjustment of industrial structure":

> ...in the period of the seventh and eighth five-year plans (1986 to 1995), we have chosen a batch of light-industry enterprises as the developmental focus, such as Meiling (making household electrical appliances), Yangtze (making household electrical appliances), Royalstar (making household electrical appliances), Gujing (making liquor) and Fangcao (making toothpaste and detergent), as this led our province toward 'a significant province of light industry' and changed the image of Anhui. If we choose the car industry as a pillar industry in the ninth five-year plan and quickly develop car production to a large scale, our province will have a brand-new advantageous sector and a new look in the twenty-first century. By then, no-one could deny Anhui's competitiveness in whatever modern sectors! (Wang 1996a: 7–8; content in brackets is added by the author)

Thoughts of establishing a local car industry also emerged at this time from the political leaders of the municipal government of Wuhu, a city of Anhui Province located on the Yangtze River. Prior to the market reform, its economy was much as that of Anhui Province in miniature, especially with regard to the lack of large industrial projects. In consequence, local industrial development was slow, and the city's industrial structure had long been biased towards light sectors. Even within the province, Wuhu was economically disadvantaged in comparison with other cities such as Maanshan and Tongling, both of which at least had a few extractive enterprises. A new door opened in 1990, when the central government made the daring decision to open up Pudong, a district of Shanghai, as an economic development zone. This immediately generated spill-over effects for many connected areas along the Yangtze River. Wuhu was soon chosen as a pivotal city for the development of Anhui, being first approved as a national-level "open city", and later warranted to build the first national-level economic and technological development zone in the province. Having been handed such an unprecedented opportunity, Wuhu officials started to develop a strong appetite for the establishment of a local automobile sector.

In this process, a few seemingly accidental events helped trigger the car dream of these officials. In 1992, several businessmen from Henan province took advantage of a local military factory to manufacture a minibus, branded Tongbao. The project soon earned a profit that greatly impressed Wuhu officials. In 1993, the local government invited FAW to jointly develop a bus-manufacturing firm within its economic and technological development zone, and the good annual sales achieved encouraged officials to pursue carmaking. Lastly, in 1995, while on a visit to Europe, a group of Wuhu officials happened upon an automobile engine assembly line that was up for sale and quickly recognized this as a golden opportunity to start their own car business. The introduction of this assembly line was a key first step towards production for Chery Auto.

For both levels of government, the plans for a local car industry were not without internal debate, though the pro-car camp ultimately prevailed. To set foot in this new field of industry was widely considered a risky move, especially for those familiar with the technological complexity of automobile production. Indeed, an interesting aside to these policy debates was that the keenest criticisms mostly came from automobile experts, whereas the ardent supporters of Anhui's car-making dream were always laymen with respect to the automobile industry (Zhang and Chen 2005). The birth of Chery Auto was thus most heavily promoted by a group of inexperienced decision makers "knowing nothing, thus fearing nothing", as recalled by people involved in this debate. In this sense, the Chery project was not the outcome of careful calculations, but more of enthusiasm and entrepreneurial vision on the part of local politicians. Relegated for decades to the status of a resource-contributing province, the local officials of Anhui had finally found an outlet for their industrialization ambitions. Theoretically, it was the institutional force, especially the cognitive factor, that mattered here, rather than "rational choice".

Industrial Concentration as the Central Logic

A basic concern of the central government in China's car sector since the 1980s was to strictly regulate any new market entrants in order to protect the nascent industry in which its hopes for national economic growth were so heavily invested. Under these guidelines, only a few major carmakers approved by the central administration were allowed to exist. The rationale behind blocking unwanted entrants and privileging certain automakers was to concentrate the industrial structure of the domestic automobile sector. National restrictions on new entrants to the car sector were formally announced in 1987, when the central government began to choose several major enterprises as "legitimate" carmakers who would spearhead the development of the industry. Under a plan introduced the following year, only the "Big Three" (FAW, SAW, and Shanghai Automobile) and "Small Three" (car enterprises in Tianjin, Beijing, and Guangzhou) received state approval, and no other automakers were allowed to operate.

Further institutional mechanisms established to enforce the restrictions were, first, an investment approval procedure to monitor domestic investments in car production and rule out any unwanted newcomers, and second, a system known as "directory management". The directory listed all automobile products approved by the central authority, while any unlisted products were prohibited from appearing in the market. For a Chinese carmaker to seek enrollment in this directory was as complex as acquiring a birth permit. The rigid regulation scheme was laid down in China's 1994 industrial policy, and dominated the Chinese automobile sector until the early 2000s (see State Council of China 1994b).

Taking the pre-reform history of the Chinese automobile industry into consideration, it is not hard to understand why the central administrators decided to regulate car production so strictly. Compared to the automotive industries of the world, China had one of the least concentrated industrial structures for automobile production at the beginning of the market reform. During the three decades of the planned economy from the 1950s to the late 1970s, the industrial structure of the Chinese automobile sector had evolved into a somewhat crowded yet scattered state. The total number of auto assemblers in China had undergone three waves of growth, and finally over one hundred assemblers were simultaneously clustered in the same market (Fig. 4.3). Moreover, by and large, most Chinese assemblers were small

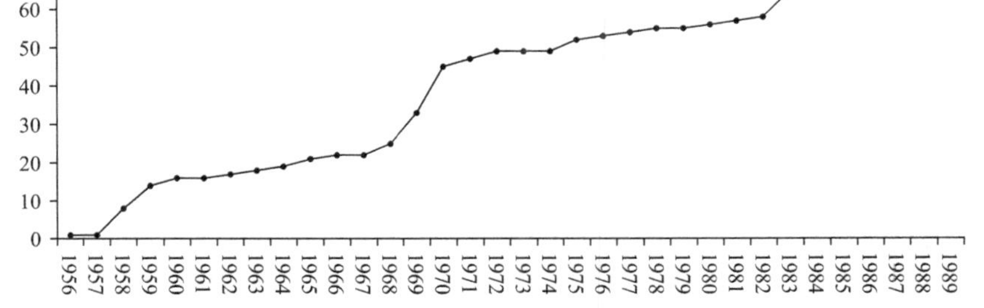

Fig. 4.3 Number of automobile assemblers in China, 1956–1989
Source: Based on China Automotive Industry Yearbooks

scale, some producing less than 100 vehicles annually. By comparison, the total number of independent automakers around the world decreased from about 50 to 30 in the 1980s (Ronald Burger Strategy Consultants 2001).

The central government had every reason to regret such a scattered industrial structure, as the formation of these small automakers had mostly been driven by periodic decentralizations within the planned economy. Under China's command economy, the central government had dominated the management of the automobile industry. In the early 1950s, the Bureau of Automobile Industry Management within the First Ministry of Mechanical Industry was first assigned as the lead institution in charge of national automobile production. But following the Great Leap Forward and its associated first wave of decentralization from 1958, this hierarchical management structure was dismantled. The central government proclaimed the idealistic desire to "walk with two legs", which for the auto sector translated to local governments developing their own automobile enterprises beyond the projects of central government. This drove the increase in local manufacturers so that by 1964 there were 19 automobile factories across China, most of which were very small. Aside from auto-giant FAW, all other automakers combined generally accounted for less than 10% of total national production. Anxieties over an overcrowded field of small manufacturers with very limited output ultimately led the central government to seize back the devolved management authority. To this end, a national corporation was organized in 1964, with the mission of integrating automobile factories through a trust. However this was soon interrupted again by the Cultural Revolution beginning in 1966, when the central administration required each province to build a complete industrial system. In this second wave of decentralization, the number of auto assemblers more than doubled, from 22 in 1966 to 45 in 1970. The largest expansion of assemblers took place in the early 1980s, however, when the new round of decentralization launched after the market reform began to take hold. When demand for automobiles started to rocket, in just three years from 1982 to 1985, the number of assemblers almost doubled again, from 58 to 114.

At the beginning of the market reform national automobile policymakers were concerned that the Chinese automobile sector comprehensively lagged behind that of advanced countries; the over-populated industrial structure with numerous small enterprises was the most significant problem. A highly centralized industrial structure was seen as a prerequisite of modern automobile manufacturing, so national policymakers proposed that the ideal structure for China's auto sector would comprise only two to three large enterprises, and six to seven medium-size enterprises (State Council of China 1994a, b). That was the rationale behind the designations "Big Three" and "Small Three". But how could such industrial concentration be realized? The somewhat paradoxical answer for central administrators in the reform context was to revert to direct bureaucratic intervention, rather than relying on the market. The lessons of the immediate post-reform decentralization rang in the ears of the central government: if left unchecked, there would be a swarm of newcomers in the sector, risking inefficient duplication, repeat construction, and waste of resources. Only bureaucratic regulation of entrants, it was felt, could channel limited national resources into a small number of automakers, thus setting up a healthier, more concentrated industrial structure.

The downside, however, was that the strict regulatory ban on unapproved automobile projects stymied local car-making initiatives, leading to great tensions between local and central authorities. The excessive policy bias towards large and established automakers such as FAW would place new, localized car projects such as Anhui's Chery Auto at an obvious disadvantage. For the auto sector, the new engine of local economic growth that the state sought to unleash through decentralization was regarded as against the national interest. Local governments had two possible responses: either surrender any banned initiative, or adopt strategies to avoid the gaze—and potential punishment—of the central authorities. As expressed by Wang Yang, there was no ideal solution for Anhui in the face of this predicament, except "to be self-reliant and struggle for growth" (1996b: 7). The clash between central and local governments also had important implications for technological strategies for non-permitted car projects. The banned status of these projects made acquiring advanced technology through joint ventures with foreign firms impossible, even though the local government may have preferred this route. Consequently, self-reliance was the only option left for Chery Auto.

Chery, from Secret Beginning to Unexpected Success

Proactive and ambitious local officials lacking car production experience and facing centrally imposed restrictions thus provided the context for the socio-political discourse of Chery Auto. Enjoying unprecedented autonomy after the market reform, the pro-car elements of the Anhui and Wuhu governments generated the momentum needed for the Chery project, which was not in line with national industrial policy. Chery Auto was nevertheless established as an enterprise owned and run by the local governments. Lack of central approval prevented it from seeking joint ventures, so technological independence turned out to be the only option. Behind the resultant production model lay a series of tense interactions between central and local authorities, as a result of which Chery Auto was to strategically cope with tight central regulations, survive, and ultimately prosper in the domestic market. The story of Chery Auto's struggle is literally reflected in its brand: *Chery* combines two Chinese characters to convey a sense of unexpectedness and final success. The first one is pronounced *che*, meaning surprise, and the second, pronounced *ry*, implies good luck—much needed by the local officials in an unfavorable policy environment that might otherwise have landed them in grave trouble (Zhang 2005).

"Hijacking" the Central Government

Linked by shared ambition, Anhui and Wuhu officials began to work closely together on the local car project in the 1990s, with Chery Auto finally established in 1997. Most of its initial investments came from five local companies owned by the Anhui or Wuhu authorities: Anhui Chuangxin Investment Co.; Anhui Investment Group Holdings Co.; Anhui Guoyuan Holding (Group) Co.; WCIC; and the Construction Corporation of Wuhu Economic and Technological Development Zone. Coordination between the two bureaucratic levels was never a major issue in practice, as there was a clear division of labor in the government hierarchy. The provincial government played the role of general director, while the Wuhu government oversaw concrete tasks such as preparing land and building infrastructure. Such a significant project required flexible and creative leadership. Zhan Xialai, the main founder of Chery Auto, had previously worked in the provincial government but was assigned in the early 1990s to Wuhu city as assistant mayor with special responsibility for developing the local car industry. The establishment and resulting success of Chery Auto in the late 1990s later smoothed the way for Zhan's promotion to secretary of the municipal committee, one of the top political positions in Wuhu city. Finally, in 2008, Zhan returned to a provincial post. In this sense, Zhan Xialai was thus not only a "businessman with red hat", but also a "floating official" between provincial and municipal levels.

However, the close coordination of local governments alone could not ensure the survival of Chery Auto under central surveillance; the key challenge for this newborn carmaker was to find a way to acquire a birth permit. Local policymakers had to be extremely careful in such a highly regulated sector, or face serious political punishment and negative economic consequences. The strategy conceived by the Anhui and Wuhu governments could be termed "hijacking the central government". They planned to secretly invest in the unapproved project first, aiming to reach a large scale before getting caught, whereupon if the central government were to discover the ploy, it would already be too late or too costly to rescind such an expensive project—in other words, it would be too big to fail. It is in this sense that the central government would be "hijacked". A crucial aspect of the strategy was that the local project must be carried out underground from the very beginning, until the investments were substantial enough that they could not feasibly be reversed. In reality, such a ruse has been deployed by other local governments across China in similar circumstances. If the project turns out to be a success, its regulation-breaking actions may often be acknowledged and even rewarded by the central government. The feasibility of "hijacking" is thus based on a tacit agreement between the central and local governments that an infraction on the part of the local government may be allowed, but all political risks should be also assumed at the local level. This unspoken agreement is an essential feature of the central–local relationship following China's market reform.

The "hijacking" strategy usually requires the local government to secretly initiate the project and rapidly make substantial progress, precisely as was the case in the Chery Auto project. In 1997, Wuhu city was selected as the location for Chery Auto, which was assigned the secret code 951 (meaning the number one project in the ninth five-year plan of Anhui Province). To hide the real intention of making cars, the project was started under a front organization, Anhui Automobile Component Industry Corporation, whose supposed focus was on auto-part production. Under this disguise, local resources were quickly mobilized to construct the infrastructure needed for production of complete cars: land totaling about 80 hectares was designated at Wuhu Economic and Technological Development Zone; and at least 1.75 billion yuan was invested by the five state-owned corporations, all of which were later requested to re-invest all annual dividends to guarantee a supply of capital for the nascent firm (Wang 2003). Construction of infrastructure was started by March 1997, and the equipment for the main stages of car manufacture were fully installed by early 1998. In May 1999, the first engine was successfully produced, and the first car, model A11, branded Fulwin, was completed by the end of 1999. Thus, in only about 30 months, Chery Auto went from a figment of (un)official imagination to producing a complete car.

Behind the remarkably rapid establishment of Chery Auto were the arduous efforts of Chery's founders, most of whom were archetypal "businessmen with red hat". Reflecting the degree of secrecy surrounding the project, the initial team comprised only eight persons, headed by Zhan Xialai. In an interview with Yin Tongyao, who was among those eight and later inherited Zhan's role as director, Yin recalled the challenges of getting Chery Auto started:

> When I first came, there was nothing but farm land. After the village moved and lands were flattened, this zone was then left to us. But there were no roads and everywhere was wilderness. We rented a few apartments behind the zone, where we did technical layout, drew blueprints, developed training programs, recruited more people, and even cooked by ourselves. It all sounds like a joke. (Chinese Central Television 2006)

Having established the production facility in secret, an even bigger challenge for the local bureaucracy than producing the cars was creating a market for them. Where would such an outlet be listed in the national directory? After all, making a car underground was possible, but selling a car without approval was unimaginable: according to national policy, these cars were not permitted to appear in the market at all. Potential consumers understandably placed very little trust in "illegal" cars and their manufacturers, hence marketing the Fulwin would be extremely hard. Moreover, due to the general lack of car-making experience and the haste of the whole project, the quality of the first batch of Chery cars was not high. In the face of these barriers, the local governments did whatever they could to fight for market space for their products. The initial solution was straightforward: the local governments would buy all the cars themselves. The Wuhu government placed the first order for the Fulwin, after Zhan Xialai issued the decision that all official-use cars in the city would be Chery products. Another initial outlet for the Fulwin was the local taxi business, over which municipal officials held much policy influence. About 400 of the earliest Chery buyers were local taxi drivers, who were rewarded with a free taxi license for eight years. This car-for-license model was later revised so that taxi drivers using a Chery would pay a reduced license fee (Wang 2003). In fact, if a license application was made for a non-Chery car, it was said to be difficult to get approval from the local taxi management agency. By 2000, Chery had produced a total of about 2000 cars, most of which became the local taxis (Lu and Feng 2005).

After the first generation of Chery cars had been produced and sold, the next step in "hijacking" the central government was to lobby for official acceptance of Chery Auto in the national directory. If this was not obtained, whatever they produced, the market for Chery Auto might be limited to Wuhu city taxis. The situation actually deteriorated around the year 2000, when China's Public Security Department issued a public notice forbidding Chery cars to be registered nationwide. Chery Auto was quickly ensnared in a crisis: even without good sales, the assembly line could not stop, otherwise component suppliers would cease to cooperate. To avoid being clearly blacklisted, a major emergency, the local governments needed to persuade central government by any and all means to accept this local "mistake". According to Wang (2003), Xu Zhongling, then governor of Anhui, traveled to Beijing to negotiate with central administrators. Seeking sympathy from above, Xu "admitted this local mistake whenever needed" (Wang 2003: 38), and strongly underscored the importance of keeping what was by now an economically significant project for underdeveloped Anhui Province. Having mobilized various social and political resources, the lobby saw positive outcomes in early 2001. Although the central agency did not directly issue a permit, a compromise was struck, later termed "getting eggs from a borrowed hen". According to this solution, in order to be listed in the national directory, Chery had to make a deal with the Shanghai Automotive Industry (Group) Corporation (SAIC), one of the "Big Three", in which Chery handed over 20% of its stock and changed its name to SAIC-Chery Auto. For Chery Auto and SAIC, the deal did not equate to a merger in any real sense, or to formal cooperation, but was merely a matter of expediency to save Chery. For the directors of SAIC, as an underground carmaker, Chery Auto might possibly prove a burden, and they were therefore unwilling to become deeply involved, though they would share permits under the arrangements set by the central administrators. For the founders of Chery Auto, to be taken over or controlled by another corporation was also not a preferred fate, as they had already progressed so far under their own steam. These mutual precautions led to the agreement of "Four Nos" between SAIC and Chery Auto as an appendix to the deal, stipulating that SAIC should not be involved in any investment, management, debt-bearing, or dividend-sharing activities of Chery Auto.

By 2004, the strategy of "hijacking the central government" had achieved unimaginable success. Freshly endowed with an official market pass from SAIC, the full potential of Chery Auto had been quickly unleashed in 2001, sales of the Fulwin reaching 28,000 units, and almost doubling in the following year. This performance enabled Chery to rapidly ascend to be one of the top carmakers in the domestic market. In 2003, Chery introduced three new models: the S11, branded QQ (a mini-car model), A15, branded Cowin (a sedan), an upgraded version of the Fulwin, and the B11, branded Easter (a luxury sedan). In the same year, sales hit 90,000 units. The achievements of Chery Auto, marked by its increasing market share, provided more bargaining chips for Chery in its negotiations with SAIC, and the temporary marriage between the two parties was eventually dissolved in 2003, Chery taking back its 20% stake. In 2004, the central government officially listed Chery as a legitimate carmaker, independent of SAIC: relaxation of the national entry regulations were in the early 2000s had greatly facilitated Chery's acquisition of legitimate status. In 2001, the old directory management model was replaced by a new scheme—announcement management—which emphasized technological standards rather than bureaucratic interventions for managing market entrants. This trend towards market governance reflected a new maturity in the Chinese automobile sector, a policy change that will be discussed in Chapter 5.

Technological Independence, a Forced Choice

Without timely technological acquisition, there would have been no possibility of Chery Auto rising up the domestic market. However, unlike FAW or SAIC, which had engaged in joint ventures for the specific purpose of technological upgrading, Chery was deprived of such opportunities by industrial policy, and was therefore institutionally compelled to be technological self-sufficient from the outset. Treading a difficult path to acquire the technologies it needed, develop own-brand models, and recruit R&D teams, Chery Auto adopted numerous tactics, some of which even led to legal wrangles. The following extracts from a media interview give a flavor of this process.

Reporter: "(Let us imagine the technology learning process to make a car is like climbing Mount Everest.) We have seen many people climbing through 'the south slope', which was not very challenging, as they could find foreign partners to set up joint ventures or cooperation. They could use foreign capital, technologies, advanced equipment, and even the best car models. While, Mr. Yin (Yin Tongyao, chief executive of Chery Auto) seems to choose a relatively harder way, going through 'the north slope' (namely, to make cars independently). Do you agree with me?"

Yin Tongyao: "Exactly. In fact, we have tried both slopes, the south and the north. The south slope, however, is not interesting to us, because on the top of that mountain will be flags of other countries, not ours. We would rather climb through 'the north slope', and plant our own flag." (Chinese Central Television 2005, content in brackets is added by the author)

Yin's claims about "the north slope" might sound stirring, but Chery's choice of technological independence cannot be simply attributed to nationalistic emotions and personal ideals. It was, rather, a result of the tension between the central and local governments, which had ruled out all other options but for Chery to produce cars on its own. For this reason, in the same interview, Yin later offered another version of how Chery "chose" to make national-brand cars, which reveals a deeper truth about Chery Auto's pathway to technological upgrading.

Reporter: "Many people wish to do self-reliant R&D, but it is too hard to proceed. Why are you so determined to follow this route?"

Yin Tongyao: "In the beginning, we did not choose, but were forced onto such a way, because there was no other way out. At that time, no one would help us. Later, when we felt this was doable, we then went ahead by ourselves......" (Chinese Central Television 2005)

Chery Auto's forced choice is further underlined by the fact that if there had been any chance of engaging a foreign partner at the beginning of the project, Anhui officials would have seized the opportunity without hesitation and Chery Auto could rapidly have enjoyed a commanding and profitable share of the domestic market. By the 1990s, in the Chinese automobile sector, a joint venture with foreign corporations was well understood as a shortcut to production technologies, advanced management experience, and capital input. Throughout the early days of Chery Auto, the possibility of joint ventures was constantly discussed in Wuhu, but all attempts failed (Wang 2003). Another possibility might have been to ally with existing large domestic automakers, such as FAW and SAIC.

But this route was no more viable than joint ventures. The Chinese "Big Three" were unlikely to show enthusiasm toward this newborn firm. Even though SAIC was later persuaded by the central authorities to hold some shares in Chery, the "Four Nos" stipulated in the deal meant that Chery could not seek any substantial technological assistance from SAIC.

Chery Auto's directors applied three upgrading strategies at different developmental stages. The initial method was *integrative imitation*, whereby new car models were based on a combination of core technologies purchased directly from foreign countries, and automobile components available in the domestic market. In developing the first Fulwin model, Chery looked to the Jetta, by then a popular product of the joint venture between FAW and Volkswagen (VW). Chery deliberately chose technologies compatible with the Jetta , mostly because it could take full advantage of the *Jetta*'s established supply system. Given that production of the Jetta had already fostered a large number of component enterprises, Chery was able to save on the investment that would have been needed to develop its own supply system. For core components such as engine and bodywork, which were not purchasable from the domestic market, Chery sought help from outside. Although less information has been officially released about the body, Seat, the largest automaker in Spain, may have been the provider for Chery. With their own production line for engines and bodywork, and the Jetta's supply system, the first Fulwin model was successfully completed within about seven months. Volkswagen, the owner of the Jetta and foreign partner of FAW, was unsurprisingly irritated by this audacious copy, and advised all its suppliers not to do business with Chery. Media reports suggest that it was SAIC who intervened to handle this issue.

Integrative imitation had enabled Chery to master the basics, freeing the enterprise to pursue a second round of technological upgrading via a strategy of *creative imitation*, attempting to independently conduct systematic whole-car R&D, albeit by imitating models already in the market. QQ and Easter, Chery's new models in 2003, were both developed in house by absorbing designs and technologies from outside. Although these imitations were never officially confirmed, the QQ and Easter appear to have been respectively modeled on the Daewoo Auto Matiz minicar and Magnus luxury sedan, produced in South Korea. Chery was criticized for "stealing" these models when Daewoo Auto went bankrupt in about 2002. However, it is not entirely fair to assert that Chery copied the Daewoo models, as independent traits were also evident in the QQ and Easter, even to the extent of them having legal patents. Whatever the extent of the replicated elements, acquiringthe capability to implement systematic whole-car R&D was a milestone for Chery Auto. But trouble was never far away. General Motors (GM), after acquiring the bankrupt Daewoo Auto, planned to introduce a Matiz-based model to the Chinese market, and soon decided to protect their intellectual property rights from infringement by Chery's QQ. GM initiated lawsuits against Chery in 2003, and the ensuing legal battle lasted about three years, ending in compromise. Defending itself from hostile action by top global automakers, Chery's technological imitation strategies had survived.

The problems that arose from creative imitation led to the need for a third strategy of technological upgrading at Chery Auto, *learning by outsourcing*, which nurtured a formal R&D capacity going beyond imitation. Chery's previous imitations were mostly based on reverse engineering the target models. In this new stage, Chery aimed to conduct R&D with no pre-existing templates. With a view to seeking help from specialized R&D firms overseas, and capturing learning opportunities, Chery began to outsource R&D to a group of world-class corporations specializing in certain aspects of automotive R&D, while trying hard to develop its own R&D capacity. While actively purchasing advanced designs and technologies from R&D firms in order to quickly improve its own models, Chery simultaneously participated in the R&D process with these firms to train its engineering team. One of the most urgent items on Chery's R&D list was to acquire its own engine, and in pursuit of this objective, in 2002 Chery entrusted the R&D of its engines to AVL Corporation, one of the world's top engine developers. Chery was allowed to send its own engineers to take part in the AVL process. With the help of AVL, by the end of 2002 Chery had developed 18 types of international-standard engines ranging from 0.8 L to 4.0 L with its own intellectual property rights, 14 of which were mainly developed by Chery's engineers (Zhan 2008). More importantly, by 2003, Chery had trained a large R&D team of about 200 technicians specialized in automobile engines (Zun 2007). These achievements had a significant effect on Chery's manufacturing capabilities, and in 2006, Chery-made engines were even exported under contract to Italy's Fiat. The same strategy was applied in other R&D areas: in whole-car design, Chery cooperated with renowned corporations such as Bertone and Pinifarina of Italy, Lotus and MIRA of Britain, and Sivax of Japan; in hybrid power, Chery worked with Ricardo of Britain; and regarding paints, Chery invited Dürr from Germany to partner it (Feng et al. 2007).

These tasks could all too easily have proved futile without effective recruitment and management of human resources, including not only R&D engineers, but technicians, managers, and qualified workers in charge of production and management. A major source of manpower for Chery was the domestic automobile enterprises, especially those which had established joint ventures. FAW-VW was one of the key targets for Chery in hunting human capital. Approximately two hundred FAW employees transferred to Chery Auto in the early years of its development. Yin Tongyao, later chief executive of Chery Auto, was one of them. Before joining Chery, he was a renowned manager at FAW-VW. How did FAW cope with losing staff on such a large scale? The answer lies in active networking by Chery's leaders, who took advantage of their *guanxi* with fellow provincials. *Guanxi*, or social ties, is a vital element of Chinese culture: connections among people from the same hometown is an important form, often mobilized for market-related purposes. At the time that the Chery founders first began to seek out human resources, the top FAW leader happened to be Gen Shaojie, who was also Anhuinese. This was a golden opportunity to put *guanxi* to work. Gen himself had been long supportive of home-grown automobile production, and had backed the bus project mentioned above as one of the early automobile plants in Wuhu city. The link with such a key figure greatly facilitated the movement of human resources from FAW to Chery: for example, on Gen's instructions FAW employees moving to Chery Auto were counted as resignations, but those who left for other corporations would be recorded as fired, with negative connotations for the departing individual (Nan and Chen 2004). It is also worth mentioning that Yin Tongyao himself was born in Anhui, making it easy for Zhan Xialai to persuade Yin to join Chery Auto.

In addition to human resources from FAW, Chery also received about twenty core technicians from DMC, one of the "Big Three", which had dissolved its R&D division in 2000 when setting up a joint venture became a strategic priority (Lu and Feng 2005). This team was to become the main force in developing the QQ, Easter, and Cowin models. One of these engineers later commented that "when the previous boss belittled us, Chery offered us another chance to prove ourselves otherwise" (Lu and Feng 2005: 69). This remark illustrates not only the different management styles of Chery and DMC, but the fact that the two automakers had chosen divergent technological upgrading paths in the same market. Chery also attracted many technicians from overseas. Xu Min, the head of Chery's R&D arm, had previously worked for Visteon Co. in the United

States, a world-class automobile components specialist; Yuan Tao, head of Chery's purchasing department, had studied for a PhD in France, specializing in automobile engines; and Gu Yu, the manager of one of Chery's holding corporations, was previously employed by Fuji Co. in Japan (Tao 2005). Chery's overseas technicians also included non-Chinese experts, prominent among whom was Terada Shinji, a management expert from Mitsubishi Motors in Japan, who joined Chery as the core technician for quality management (Zun 2007).

Above all else, it was the acquisition of technology and human resources that dramatically accelerated Chery Auto towards technological independence. This is demonstrated by the number of patents it obtained. As can be seen in Fig. 4.4, Chery ranked fifth among all automakers in China in terms of patents acquired in 2006, even though it was less than ten years old and the technological upgrading was in its infancy at the time. More surprisingly, Chery surpassed all domestic automakers, including the "Big Three". Technological independence was an obvious reason for Chery to be active in acquiring patents, whereas other domestic automakers with

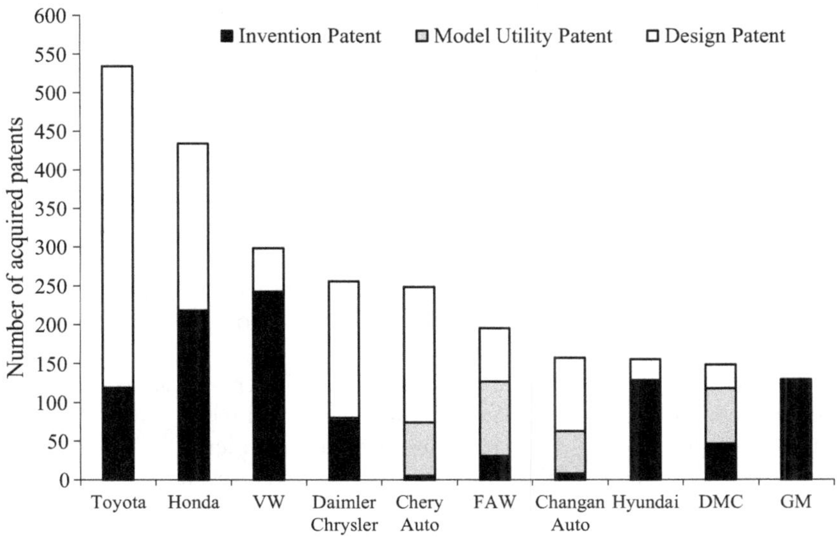

Fig. 4.4 Top ten automakers in China in terms of the number of patents acquired in 2006
Source: Based on Zhao 2006

access to foreign technologies through joint ventures were less motivated to do so. Figure 4.4 shows statistics subdivided by three types of patent, as defined by Chinese patent law: invention patents, model utilities patents, and design patents. As can be seen, Chery's patents were mostly for model utilities or designs, which are less significant than invention patents in terms of R&D standards. By contrast, the foreign corporations listed all had much larger number of invention patents.

One critical question remains unanswered: how did Chery so successfully gain a commanding share of the domestic market? Though the work of local officials towards obtaining a permit and acquiring technologies is important, the unique market advantage of Chery cars was their low price. To carve out a space in a market filled with mature models mostly from joint ventures, Chery had to compete at as low a price band as possible. Its first model, the Fulwin, for instance was sold at only about US$ 10,000, 30–40% cheaper than most comparable models in the market at the time.

How was it possible to operate at such a low price? The secret can be traced once again back to the unique production model of Chery Auto. Born into an unfavorable or even discriminatory environment, the Chery Auto project had access to minimal financial resources, and cost savings were thus crucial to its survival. For example, most infrastructure designs were done by Chery itself, rather than being contracted to outside companies, saving arge sums. The enterprise also operated on a rolling development principle, whereby infrastructure and equipment were not completed all at once, but gradually expanded according to actual demand when production increased. As a result, the combined costs of infrastructure construction for Chery Auto were much lower than the projections (Chen 2008). The total investment in the entire Chery project was less than 2.0 billion yuan, significantly lower than investments in joint venture projects such as FAW-VW (about 14.0 billion yuan) and Dongfeng-Citroën (about 13.1 billion yuan) (Li and Niu 2003).

Moreover, "Chery was standing on the shoulders of giants, but not held in the hands of giants" (Yang 2008). That is, the low price of Chery Auto was also down to its technological independence. Chery could freely take advantage of the domestic supply system established by joint ventures, choose the best prices in the market to acquire the components it needed, and thus drive down the final cost hard. More importantly, technological independence also helped Chery to avoid forms of rent-seeking by foreign corporations in joint ventures. For instance, foreign corporations often made additional charges to Chinese partners for model introduction, equipment installation, and technical support. As a result, while the usual cost of introducing a new car model was about 2.5–4.0 million yuan in the market, joint ventures often had to pay ten to twenty times this to their foreign partners for similar designs (Fu and Zhang 2006). In this way, a significant portion of the profits were indirectly transferred to foreign partners, and in turn, these extra charges greatly increased the final costs to the consumer. Chery, as a technologically independent corporation, had no such middleman fees to pay.

As an early manufacturer of national cars after the market reform, Chery Auto has continued to perform well in recent years. Its cumulative production totaled 1 million vehicles in 2007, 2 million in 2010, 3 million in 2013, and 5 million in 2015, respectively. Among national car manufacturers, Chery is the first to hit these records, though still behind the established joint ventures. However, unlike the joint ventures which concentrated on the domestic market, Chery Auto is aggressive in the international market. Since 2001, Chery Auto has been the leading car exporter in China, reaching markets in the Middle East, Southeast Asia, Eastern Europe, and other regions. From 2007 to 2010, more than 200,000 cars were exported by Chery, accounting for almost half of China's total car exports. The company has also established overseas plants to facilitate its global reach. Both the significant increase in production and its vigorous international expansion affirm the status of Chery as a champion producer of national cars in the Chinese automobile sector in the early 2000s.

Case Summary

Chery Auto represents the third production model, in which an automobile enterprise belonging to the local government manufactures national-brand cars without a foreign joint venture partner. This chapter tells a story of conflict, compromise, and struggle between the local and central authorities, in which the Anhui government's bid to promote local car production was initially unacceptable under the policy directions of central administrators. Based on the institutional framework proposed in this book, Chery Auto's car project and the associated local–central tensions can be attributed to the pre-reform economic designation of Anhui as a supporting region for the industrialization of other prioritized areas, tasked with contributing raw materials and agricultural products. The path-dependency effect in Anhui resulted from long-term neglect by central planners, which inadvertently cultivated a remarkable drive towards industrialization and especially car manufacturing among local officials, who lost no time in acting on their ambitions once market reform opened the door for local governments to pursue their own economic goals. The car-making vision of the local government was officially denied by the central authority, however, since new market entrants were considered a threat to the industrial concentration of the entire national automotive sector. This chapter has thus demonstrated a scenario in which prior economic positioning drove a discursive wedge between the central and local authorities, exposing the local car project to the full force of ideological tensions. The chapter has further explored the resulting socio-political process by which Chery Auto was able to establish itself, as proactive local officials struggled against the central will. As shown in this chapter, Chery had to be initiated as a local secret without notifying the central authority, and rapidly developed to the point that it was already too late to be rescinded once discovered by the central authority. The fact that Chery had to "hijack" the regulatory environment in this way meant that technological upgrading could only be achieved independently, as central restrictions forced the project underground with no possibility of engaging foreign partners. The tortuous route depicted in this chapter finally shaped the production model of Chery Auto, a local state-owned enterprise that would go on to champion independent national brands in the Chinese automobile sector.

Notes

1. One jin is equal to half a kilogram.
2. Developed as a part of the Great Leap Forward movement (1958–1962), the People' Communes were rural organizations in China for collectivization of agricultural production.

Bibliography

An, H., & Ren, Z. (2008). Reconsideration of China's Traditional Planned Economic System from the Perspective of Twice Division of Government Power in the Past. *Journal of Sichuan University (Social Science), 2*, 95–101. (in Chinese).

Bao, X. (1992). Developing Characteristic Products Is an Important Way of Invigorating Machinery Industry. *Economic Management, 8*, 16–17. (in Chinese).

Chen, Y. (2008). Why Chery Helds Enthusiasm for Innovation. *China New Time, 6*, 86–89. (in Chinese).
Chinese Central Television. (2005). *Chery Contributes to the Competitiveness of the Chinese Automobile Industry.* http://www.cctv.com/program/dialogue/20050309/101611.shtml. Accessed 10 June 2017. (in Chinese).
Chinese Central Television. (2006). *Oriental Son: Yin Tongyao, Director of Chery Auto.* http://news.sina.com.cn/s/2006-06-29/212710289386.shtml. Accessed 10 June 2017. (in Chinese).
Economic and Trade Commission of Anhui Province. (1999). The Great Achievement of Industrial Development. In the Anhui Government (Ed.), *Fifty Years of Anhui* (pp. 329–334). Beijing: China Statistics Press. (in Chinese).
Editorial Committee of Anhui Yearbook. (2000). *Anhui Yearbook 1999.* Hefei: Anhui Yearbook Press. (in Chinese).
Fan, L. (2003). Businessman with Red Hat in Anhui. *Decision and Information, 10*, 19–20. (in Chinese).
Feng, K., Yin, T., & Wang, Y. (2007). The Innovation Mode of Chery. *China Soft Science, 3*, 76–84. (in Chinese).
Fu, G., & Zhang, R. (2006). Self-Reliant Innovation: Strategy Under the Weakened Industry Barriers. *Contemporary Finance & Economics, 9*, 79–83. (in Chinese).
Gao, Z. (1986). Exploring the Industrial Structure of Anhui. *Theory Research, 2*, 16–22. (in Chinese).
Ge, S., & Xia, S. (1999). The Fifty Year Transformation and Characteristics of the Industrial Structure in Anhui. *Jianghuai Forum, 6*, 30–33. (In Chinese).
Hu, C., Chen, X., & Yu, X. (2001). A Comparative Study on the Economic Development of Anhui. *Shanghai Economy Review, 10*, 70–79. (in Chinese).
Jing, C. (2008). Reflecting the Ownership Institution in China from the Perspective of Central-Local Relationship and State-Owned Assets. *Lanzhou Study Journal, 5*, 98–106. (in Chinese).
Li, J., & Niu, M. (2003). Chery, a Black Horse in the Chinese Automobile Industry. *Anhui Taxation, 2*, 16–19. (in Chinese).
Liu, Y., & Zhou, D. (1983). Historical Overview of Anhui's 'Responsibility System' in 1961. *Theoretical Frontline, 179*, 1–11. (in Chinese).
Lu, F., & Feng, K. (2005). *The Policy Choice to Develop Chinese Automobile Industry with Independent Intellectual Property Rights.* Beijing: Peking University Press. (in Chinese).
Nan, Y., & Chen, P. (2004, February). Chery Running Through a Narrow Path. *Money China*, 28–33. (in Chinese).
National Planning Committee. (1997). *The Developmental Strategy of Passenger Cars for Chinese Households.* National Planning Committee. (in Chinese).

Nee, V. (1992). Organizational Dynamics of Market Transition: Hybrid Forms, Property Rights, and Mixed Economy in China. *Administrative Science Quarterly, 37*(1), 1–27.

Oi, J. C. (1992). Fiscal Reform and the Economic Foundations of Local State Corporation in China. *World Politics, 45*(1), 99–126.

Ronald Burger Strategy Consultants. (2001). *Ten Trends of the Development of China Automobile Industry After WTO*. Ronald Burger Strategy Consultants. (in Chinese).

Social Science Institute of Anhui Province. (1999). Fifty Years, Great Achievement. In Anhui Government (Ed.), *Fifty Years of Anhui* (pp. 3–15). Beijing: China Statistics Press. (in Chinese).

State Council of China. (1994a). *Outlines of the Industry Policy in the 1990s*. State Council of China. (in Chinese).

State Council of China. (1994b). *The Industry Policy of the Automobile Industry*. State Council of China. (in Chinese).

Su, H., & Ou, Y. (1987). *Anhui Developmental Strategy*. Anhui: Anhui People's Press. (in Chinese).

Tao, J. (2005). Chery, Asking Profit from Human Resources. *Chinese Times, 1*, 32–35. (in Chinese).

Wang, L. (1989). *The Discourse of Responsibility System in Anhui: 1961 to 1978*. Beijing: Kunlun Press. (in Chinese).

Wang, Y. (1996a). The Car Industry: Driving Anhui to the New Century (First Part). *Decision Reference, 2*, 6–8. (in Chinese).

Wang, Y. (1996b). The Car Industry: Driving Anhui to the New Century (Second Part). *Decision Reference, 3*, 6–8. (in Chinese).

Wang, X. (2003). Chery, the State Engine and the Market Path. *New Economy, 9*, 38–41. (in Chinese).

Wu, Y. (2008, October). *Fiscal Equalization will Help Prevent Corruptions*. Oriental Outlook. http://news.sohu.com/20081028/n260297950.shtml. Accessed 10 June 2019. (in Chinese).

Xu, C. (1986). Thoughts on the Developmental Strategy of the Economy of Anhui. *Theory Construction, 2*, 23–30. (In Chinese).

Xu, J. (2004). Businessman with Red Hat of Anhui in Trouble. *South Reviews, 4*, 53–55. (in Chinese).

Yang, M. (2008). The Legend of Chery. *Decision-Making, 9*, 21–23. (in Chinese).

Zhan, Y. (2008). An Analysis on the Dialectic Relationship Between Strategic Ally and Enterprise Innovation. *Studies in Dialectics of Nature, 24*(3), 68–72. (in Chinese).

Zhang, Z. (2005). *Top Background Information: Significant Person in the Chinese Economy*. Shantou: Shantou University Press. (in Chinese).

Zhang, S., & Chen, J. (2005). *Chery Auto Breaks the Myth: A Forced Path of Self-reliant R&D*. http://ah.anhuinews.com/system/2005/11/16/001391506.shtml. Accessed 10 June 2017. (in Chinese).

Zhao, R. (1995). Five Advantages Sectors in the Machinery Industry of Anhui. *China Electronics and Machinery Industry, 10*, 15. (in Chinese).

Zhao, X. (2006). Observing Chery from Outside: Analyze the Development of Chery from a Perspective of Patent. *China Invention & Patent, 4*, 52–55. (in Chinese).

Zhao, X., & Zhang, L. (1999). Decentralization Reforms and Regionalism in China: A Review. *International Regional Science Review, 22*(3), 251–281.

Zhou, Z. (1998). Explore the Developmental Advantages of Anhui Economy. *Economic Theory and Economic Management, 3*, 61–65. (in Chinese).

Zhu, Y. (2004). *An Investigation of Businessman with Red Hat in Wuhu*. http://news.xinhuanet.com/report/2004-02/22/content_3181726.htm. Accessed 10 June 2017. (in Chinese).

Zun, L. (2007). *Innovation of Chery*. Beijing: China Citic Press. (in Chinese).

CHAPTER 5

Geely Auto, Pioneer Grassroots Automaker

Zhejiang Geely Holding Group, abbreviated to Geely Auto, is one of the earliest fully private-owned carmakers in China. From its beginnings as a refrigerator manufacturer in the 1980s, Geely had become one of China's leading car producers by the turn of the millennium. As a private corporation, Geely Auto is unique in comparison with the previous three production models, featuring far less involvement by both central and local governments, compensated by the proactive entrepreneurship of its founders. As with each of the previous cases, Geely's path to prosperity can be traced to the status assigned to the firm's home province of Zhejiang by the communist regime. Zhejiang was largely ignored by central policymakers throughout the planned-economy era, an oversight which ironically fostered the most vibrant private economy in China immediately after the market reform. Li Shufu, the founder of Geely Auto, was among numerous private entrepreneurs in Zhejiang Province battling to carve out new markets from institutional neglect. Though entrepreneurship is the central theme in this case, the state exerted significant influence as a regulatory force. As with Chery Auto, the central government initially stood in the way of Geely's ambitions to produce cars, pursuing its industrial policy seeking to protect the concentration of the auto sector. For Geely, like Chery, these policy constraints implied a survival strategy of technological independence. Unlike Chery, however, Geely received negligible support from local state officials, in many ways navigating a tougher and riskier development pathway. The case of Geely Auto reveals how the

Q. Feng, *Variety of Development*,
https://doi.org/10.1007/978-981-10-5912-4_5

fourth production model, in which fully private firms make national-brand cars, was made possible in post-reform China.

This chapter is organized in four sections. First, the historical origin of Geely Auto is examined via the pre-reform status of Zhejiang Province on the national economic map. The chapter then explores the relationship between lack of central attention to Zhejiang and the vibrant local economy following the market reform, highlighting the active private entrepreneurs and the laissez-faire position of the local political structure. The next section focuses on the unique development ideology in Geely Auto's hometown of Taizhou city, revealing how the local machinery-focused economy and its affiliated mindsets later fostered Geely's car manufacturing ambitions. As in Chery's case, this section reveals substantial tensions in a configuration of local development ideas where spontaneous carmaking ideas encountered national regulations. Finally, the chapter analyzes the development of Geely as a socio-political process, with the firm's struggles for survival in the face of a biased central administration and indifferent local government finally leading to a unique production model entailing full private ownership and a self-sufficient technological strategy.

Zhejiang, a Forgotten Land

As a private carmaker, Geely Auto's success hinged on the determined and idealistic entrepreneurship of its founder, Li Shufu. However, it would be an oversimplification to attribute the rise of Geely solely to the pioneering character of one individual. Li was one among many thousands of like-minded businessmen in his home province of Zhejiang whose entrepreneurial energy drove the prosperity of the best private economy of China during the market transition. The origins of this large cohort of private businesspeople in the context of the high watermark of Chinese communism can be traced to the economic status of Zhejiang Province before reform. By comparison with the other locations explored in this book, Zhejiang was thoroughly neglected and forgotten in China's national economic planning agenda. Such was the extent of its disadvantage that Zhejiang achieved the dubious notoriety of "Three Nos": no natural endowment, no national investment, and no special policy treatment. In the planned-economy system, where the destiny of a local economy was mostly determined by the central authority, the prospects of Zhejiang appeared dim.

However, the economic positioning of Zhejiang as irrelevant was well justified from the perspective of national policymakers. Zhejiang lacked any significant natural resources that could contribute to national industrialization. According to national indices of basic natural resources per capita, Zhejiang ranked third from last in China (Gao 1987; Zhu 1988), which made it hard for the province to attract central investments in industrialization. Moreover, the rugged topography of Zhejiang Province, termed "seven mountains, one water, and two lands", was unsuitable for most priority economic activities. Arable land per capita was less than half the national average, and the mountainous landscape was also troublesome for transportation, a fundamental consideration in how economic and industrial activities were prioritized. Despite this difficult physical environment, Zhejiang was also highly populous: by the end of 1981, the provincial population was about 39 million, the tenth largest of all Chinese provinces, with the fourth highest population density in China (National Statistics Bureau 1982). In the face of such population pressures, Zhejiang's limited resources could barely support the development of either agricultural or industrial sectors. A further issue was Zhejiang's sensitive location, in close proximity to Taiwan on China's southeastern coast. China's long-held ambitions to reunify Taiwan placed Zhejiang Province on the frontline of any potential conflict in the planned-economy period, and these military concerns made it unthinkable for planners to locate significant industrial projects in Zhejiang.

In consequence, this province was marginalized in the era of planned economy. The province bore no similarities with the northeast, which had been chosen as the focal area for heavy industry; nor was it comparable to Shanghai, as the national economic center; and it was even less important than neighboring Anhui province, which was at least considered a significant provider of raw materials for the national economy. Unsurprisingly, central investments in Zhejiang Province were extremely meager. From 1952 to 1978, the investment by the state-owned units in Zhejiang was half the national average and the lowest of all Chinese provinces (Fang 2000). Investments in basic infrastructure in Zhejiang from 1952 to 1979 accounted for less than 2% of the national total. This situation did not change much after market reform. From 1979 to 1992, the national investment of the state-owned working unit per capita in Zhejiang reached only three-quarters of the national average, ranking 22nd among all provinces (Fang 2000).

The heavily disadvantaged status of Zhejiang during the planned economy generated an interesting byproduct in the form of underground private local business. With minimal central investment or natural endowment, the provincial economy was barely able to sustain the basic survival of its population, and for many Zhejiangnese it was necessary to seek alternative ways to make a living beyond that which the planned economy had failed to provide. Meanwhile, lack of attention from the central authority also weakened local organizations and institutions tasked with enforcing the planned economy. Supervision and monitoring of economic activities was thus less strict in Zhejiang than in many other provinces. The sheer necessity of survival and lax regulatory environment jointly contributed to the emergence of underground businesses in Zhejiang even before the market reform. At the point of instigation of the reform policies in 1978, state-owned industrial enterprises in Zhejiang only accounted for about 60% of local industrial output, far below the national average of 80%. By contrast, the proportion accounted for by non-state-owned industrial enterprises was twice the national average (Fang 2000). This directly contradicted political and ideological opposition to private business under the planned economy, which had been consolidated after the People's Communes collectivization movement beginning in the late 1950s.

In fact, Zhejiang's clandestine market never ceased functioning during the pre-reform era, even during the Cultural Revolution, when Chinese communist ideology and zeal reached their zenith. The small Zhejiang county of Yiwu offers a useful example of this phenomenon. Local peasants in Yiwu county conducted a historically embedded form of trade known as "exchanging feathers for sugar", which could be traced back to the Ming dynasty. During slack periods of the agricultural cycle, Yiwu peasants would carry brown sugar, a main local agricultural product, to other regions of China in exchange for poultry feathers, which they would later apply to their fields as fertilizer. While this curious tradition was interrupted under the communist regime, it was never eradicated: in the hardest times of the planned economy, Yiwu people found a way to practice the "feathers for sugar" exchange to make a living. Recognizing the elementary comparative advantage this system offered in otherwise bleak circumstances, it was not only tolerated by local officials, but was even incorporated into the collective production system, as these reports attest:

> At the beginning of the Cultural Revolution, even cadres could not save themselves, and ordinary peasants could not survive through the collective economy. Young people started to think of alternative ways to making a living, and I was one of them. In order to save food for my family, I went out to Zhuji and Jiangzao with a few elders of the same village to exchange feathers for sugar. (Chen 1997: 177)

> When I was 14 years old, my village had totally fifty persons of 'shoulder poles' for the businessmen of feathers for sugar. According to their agreement with the production team, they would turn in 100 jins of feathers per month, or 1 yuan per day (one jin is equal to half a kilogram). If the task was fulfilled, the team would offer them a full load of working points for a year. In fact, even with full working points, we could only be paid by 90 yuan, 400 jins of paddy, and 100 jins of wheat at the end of the year. With these foodstuffs, we could hardly survive, not even to mention any surplus. But these people going out for business could gain all working points, and also had some extra profits after turn-in. Needless to say, they were much better off than those who stayed at the production team. (Pan 1997: 102)

The case of Yiwu demonstrates the conflicting logic that a local private economy could be sustained by the unfavorable economic positioning of Zhejiang in the planned economy. Scaled up across the province, the loose control over the planned economy in Zhejiang made the operation and expansion of the private economy relatively easy. Put in the stronger words of local officials, the rise of the private economy in the province occurred simply because otherwise the people of Zhejiang had nothing at all. Economic discrimination against Zhejiang thus actually prepared the first generation of local private businessmen to gain an advantage in the upheavals of the impending market reform.

The "Zhejiang Phenomenon"

In the annals of the market transition, Zhejiang is often presented as something of an economic miracle. Chinese media reports about post-reform Zhejiang are peppered with fantastical terms such as the "Zhejiang phenomenon", "Zhejiang experience", and even "Zhejiang model". Positive assessments of Zhejiang's economic performance after the economic reform were not confined to China. In an issue of *Business Week*, American economist Huang (2008) strongly praised Zhejiang as "a free-market success story", and admonished the policymakers of China to "look no further than its own star province" when working out the nation's economic future.

The contrast between pre- and post-reform Zhejiang is stark when viewed through basic economic parameters. From 1953 to 1978, the annual GDP growth rate of Zhejiang Province was 5.7%, below the national average of 6%, but this doubled to 13.2% for 1978 to 2005, much higher than the national average of around 8% (Zhejiang Statistics Bureau 2006). In consequence, GDP in Zhejiang grew about 30-fold from 1978

Table 5.1 A comparison of Zhejiang, Guangdong, Jiangsu, and Shandong provinces in 2006

	Guangdong	*Jiangsu*	*Shandong*	*Zhejiang*
Population (1 million)	93.0	75.5	93.1	49.8
Area (1000 square kilometers)	180	100	160	100
GDP (1 billion yuan)	2596.9	2154.8	2184.7	1564.9
GDP per capita (yuan)	28,072	28,685	23,546	31,684
Disposable income per capita of urban residents (yuan)	16,016	14,084	12,192	18,265
Net income per capita of rural residents (yuan)	5080	5813	4368	7335
Total import and export value (1 billion US dollars)	527.2	160.4	95.3	139.1
Actually utilized foreign investment (1 billion US dollars)	14.5	17.4	10.0	8.9
Contribution of private economy to GDP (%)	40.0	35.0	40.2	54.9

Source: Based on Ding (2007) and Zhejiang Statistics Bureau (2007)

to the early 2000s. Accordingly, Zhejiang's position in the national economic rankings rose from 12th to 14th in the early 1980s, to the top 5 in the early 2000s. More importantly, in 2005, Zhejiang became the first province in China with per capita GDP above US$ 3000.

The "Zhejiang phenomenon" went beyond rapid economic growth alone, however, as there were many other examples of rapidly developing provinces in China during the same period. To effectively reveal the nature of the Zhejiang model requires more detailed comparison. Table 5.1 lists main economic indicators for Guangdong, Shandong, Jiangsu, and Zhejiang, China's top four provinces in terms of GDP by the year 2006. Although all four provinces had shown excellent economic performance, Zhejiang had some unique characteristics. First, Zhejiang is smaller both in population and geographical area, which may partially explain the gap in the sheer size of GDP between Zhejiang and the others. However, when we examine GDP per capita, Zhejiang becomes the strongest, achieving higher rates than Guangdong, Jiangsu, and Shandong by 12.9%, 10.5%, and 34.6%, respectively. The contrast is even more significant when it comes to per-capita income: at the time of the data, Zhejiangnese citizens were beyond doubt the richest on average in China. Compared with Guangdong, home to the second-richest per-capita income, urban residents of Zhejiang had 14% more disposable income, while the rural residents had a 44% higher net income.

The Zhejiang phenomenon can be further isolated by comparing the structure of trade and investment (Table 5.1). After the reform, a routine developmental strategy among all Chinese provinces was to take advantage of cheap domestic labor, introduce foreign investment, and promote large-scale exports. The theory was that local economic growth should be primarily driven by engaging the international division of labor. Guangdong, China's first province to open to foreign capital, represents this approach. As can be seen in Table 5.1, total imports and exports in Guangdong in 2006 were higher than the sum of the other three provinces. Though not comparable with Guangdong in foreign trade, Jiangsu led the top four in foreign investment. The development of both Guangdong and Jiangsu relied on cheap local labor, attracting foreign capital and promoting exports. As the labor price is fundamental to this development pathway, local wages could not rise easily, even in the context of rapid local growth. Zhejiang was obviously on an alternative track that was much less strongly driven by foreign trade and investment. Foreign investment actually used in Zhejiang was only about half of the figure for Jiangsu, and its foreign trade was about one-fourth that of Guangdong, but the contribution of the private economy in Zhejiang reached about 55%, much higher than the other three provinces.

The secret behind the Zhejiang phenomenon was having the most active local private economy in China at the point of the market reform. From 1978 to 2006, the Zhejiang private economy grew annually by a remarkable 28.9%, more than double the local GDP growth rate. As a result, the proportion of local GDP accounted for by the private economy surged from 5.7% in 1978 to 53.8% in 2003 (Zhejiang Statistics Bureau 2007). For eight consecutive years from 1998 to 2006, privately owned enterprises in Zhejiang were ranked number one across China in terms of registered capital, industrial output achieved, sales, total retail sales of consumer goods, foreign currency from exports, and total number of public corporations (Zhang and Yang 2006). By 2006, 203 of the top 500 Chinese private enterprises were from Zhejiang (Zhejiang Statistics Bureau 2007). This dramatic acceleration of the private economy accounts for Zhejiang's economic miracle after the reform. Unlike the development model of the other top-performing provinces, which emphasized harnessing cheap labor, the Zhejiang model brought more local benefit through higher rates of personal income.

The booming post-reform private economy in Zhejiang clearly originated in the period of central planning. In a transitional economy, market actors were not going to spontaneously arise once market rules had replaced the entrenched command-economy system. Private entrepreneurs were neither given nor automatically born out of the institutional changes associated with the reform. The seeds of the private economy in Zhejiang were sown by the disadvantaged economic position of the province before the reform: those Zhejiangnese who had operated underground businesses or aspired to do so soon became the first generation of private entrepreneurs once their visions and capabilities had been unleashed by the market. The same mechanism might not have been applicable in the many other regions of China that had enjoyed more privileged economic positions in the central planning era. For example, for individuals where well-established planned-economy institutions existed such as in Changchun or Shanghai, it would be harder to leave their established careers in such organizations as state-owned factories in favor of new private business opportunities. Even for provinces like Anhui, where local industrialization had received less central attention, for the most part people were slower to develop the degree of Zhejiangnese entrepreneurial enthusiasm to pioneer opportunities in the new market, and many remained attached to Anhui's agricultural traditions.

In many ways, the circumstances of Anhui and Zhejiang are fairly comparable. Both are located close to Shanghai, the economic center of China, and both were underdeveloped provinces by comparison with others at the beginning of the reform. However, by 2005, Zhejiang's GDP had risen to 4th in the national rankings due to its prosperous private economy, whereas Anhui had declined to 15th. This outcome plays against the perceived role of natural endowment, since Anhui was much better endowed than Zhejiang in terms of resources and human capital. What was missing in Anhui, however, was the sheer number and determination of private entrepreneurs. In Zhejiang these were the agents who initiated, catalyzed, and drove the new market forwards, emerging as an unexpected outcome of the specific historical and institutional pathways in the province, rather than a consequence of official interventions. More ironically, in consideration of the "Three Nos" attached to Zhejiang in the pre-reform period, entrepreneurship only seemed to emerge in the *absence* of state interventions. It thus seems that the institutional changes under which the market transition unfolded could occur more freely and efficiently in areas marginalized under the previous institutional structure.

It would, however, be inaccurate to deny the role of local governments as facilitators of the private economy in Zhejiang. Long overlooked by the communist regime, local officials sympathized to some extent with local "non-socialist" economic activities. However, at the beginning of the reform, to support the private economy was still politically risky, and sometimes the local government had no choice but to suppress local private entrepreneurs. For example, in the early 1980s, when the earliest post-reform private economy of Zhejiang Province began to take shape in Wenzhou city, the local government jailed several local private entrepreneurs for "speculation and profiteering". The real reason was purely political: individuals making money from the market were capitalists, and therefore at odds with mainstream socialist ideology. This negative political atmosphere gradually shifted as the reform took hold. From the 1980s to the 1990s, local governments in Zhejiang issued a series of policy documents deregulating and encouraging the private economy (Table 5.2). As these documents show, in the 1980s the key concern was to clear political

Table 5.2 Major political documents in Zhejiang supporting the local private economy, 1980s–1990s

Time	*Political document*	*Key content*
1981	Notice about developing diversified business in farming in rural areas	To encourage private business among peasants beyond traditional farming
1985	Speech by Shen Zulun, associate governor of Zhejiang, at a symposium about household industry	To argue against the prejudice towards the private economy and promote development of household-based industrial production
1985	Temporary regulations to protect the legal rights of specialized rural households	To protect rural households in developing private production and business
1989	Speech by Li Zeming, provincial party secretary of Zhejiang, at Wenzhou city	To justify the political suitability of the local private economy
1991	Notice on strengthening management of individual industrial and commercial households and private enterprises	To encourage the development of the local private economy
1992	Speech by Li Zeming, provincial party secretary of Zhejiang, at conference on individual industrial and commercial households and private enterprises	To harness the private economy as the engine of development for undeveloped areas in Zhejiang
1993	Notice about promoting the healthy development of the individual and private economy	To allow the private economy to engage in any legal business
1998	Notice about fully developing the non-collectively-owned economy	To fully develop the private economy and to create good environments for market competition

Source: Based on Fang (2000)

barriers to the private economy. A local slogan of the time proclaimed "to emphasize development regardless of (socialist or capitalist) character, and to focus on economic outcomes regardless of (socialist or capitalist) proportion". This clearly expressed the desire of the local governments to rid themselves of socialist ideological constraints, which criticized the private economy for its non-socialist "character" and thus strictly regulated its "proportion" in the overall economy.

"Taizhou Manufacture" and "the Automobile Maniac"

Not only did the disadvantaged economic positioning of Zhejiang foster the local private economy, but it was also linked to the consolidation of car-making ambitions among some local entrepreneurs. Among businessmen in Zhejiang, Li Shufu, the founder of Geely Auto, was an outlier, since unlike him, most other local private entrepreneurs concentrated on low-tech and low-investment sectors. Li's enthusiasm for producing cars seemed unique, and the Chinese media even gave him a semi-sarcastic nickname: "the automobile maniac". However, behind Geely Auto was not only Li's personal compulsion, but the distinct industrial culture of Li's small hometown of Taizhou. Similar to Chery Auto, local enthusiasm for cars was not welcomed by national industrial policy. Regulatory conflicts with the national administration ultimately led Geely Auto towards a path of technological independence, and prefaced a hard road to success for this fledgling project.

Machine Making as Industrial Culture

Most private enterprises in Zhejiang were small scale at the beginning of the reform, concentrating on light industry. Local entrepreneurs especially favored sectors such as apparel and food processing, which required less investment and technology, and were ideal for those starting out in business. So-called "small commodities" soon became Zhejiang's major industrial products, with specialized industrial clusters developing throughout cities and counties of the province, including cigarette lighters, fasteners, and spectacles in Wenzhou, neckties in Shengzhou, and stockings in Zhuji. These Zhejiang products gained a significant share of the domestic market and were later exported overseas. Accompanying the production of small commodities, specialized markets also appeared across Zhejiang, which proved a critical platform facilitating production, sales, information exchange, and technological upgrading of local products. In Yiwu, once the focus of the "feathers for sugar" trade described above, a flourishing local specialized market eventually resulted in the establishment of a world-renowned trade center for small commodities.

While Taizhou, the birthplace of Geely Auto, can be considered generally representative of the Zhejiang model, the city took a different course when it came to specialization. Surrounded by regions specializing in the production and trading of small commodities, Taizhou entrepreneurs opted to focus on manufacturing machinery. This local preference later gained public recognition as a unique industrial culture termed "Taizhou Manufacture". Prior to the mid-1990s, the food and apparel/leather industries had been the leading sectors of the Taizhou economy, and of other regions of Zhejiang. Taizhounese businesses initially favored exploiting local agriculture products such as fruits and aquatic produce, while later labor-intensive textile-related production became another advantageous local field for private enterprises. After the mid-1990s, however, these traditional sectors were gradually replaced by an increasing specialization in machinery. For example, the motorcycle industry in Taizhou became one of the largest in China, and local motorcycle component manufacture occupied about one-quarter of total national output. Taizhou sewing-machine production also became globally significant, its industrial zigzag machines eventually gaining more than half of the world market share. At the same time, Taizhou had the largest production base for valves and water pumps in China, while freezers made in Taizhou accounted for one-quarter of the domestic market. Alongside these neighboring industries, the local automobile sector, represented by Geely Auto, established itself as the new arm of Taizhou Manufacture.

The strong focus on machinery manufacture had important outcomes for the economy of Taizhou, shaping the local industrial structure. As seen in Fig. 5.1, driven by the expansion of machinery production, secondary industry began to lead local output by the mid-1990s. The flourishing machinery sector helped the local economy in Taizhou to grow more rapidly than other regions of Zhejiang Province. From 1990 to 1995, Taizhou's GDP increased annually by 24.8%, as compared to the Zhejiang average of 18.8% and national average of 12.1% (Chen and Dai 1996). Machinery production in Taizhou also spawned a number of large corporations of national renown, particularly from the mid-1990s, including Qianjiang Motor (manufacturing motorcycles), Feiyue Group (manufacturing sewing

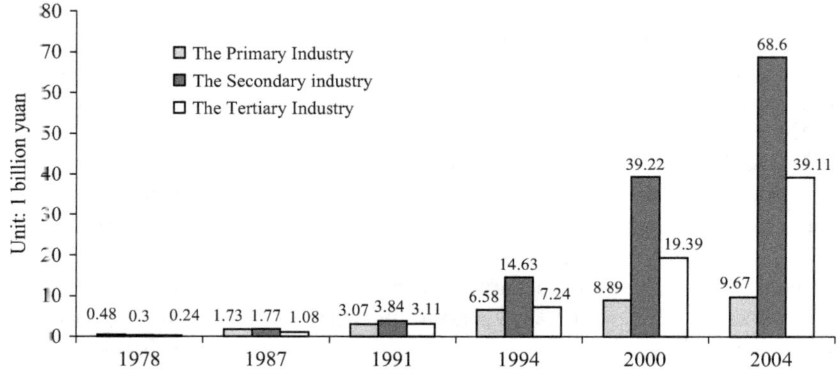

Fig. 5.1 Industrial structure of Taizhou, 1978–2004
Source: Zhang et al. 2005

machines), and Star Group (manufacturing home appliances). Although numerous small and medium-sized enterprises remained in Taizhou, large corporations gradually came to absorb smaller ones and form local industrial clusters. Finally, specialized markets for machinery-related products began to appear in Taizhou, such as the Zhejiang Material Recycling Market, Luqiao Machine-Electronics Hardware Market, Auto-Motor Component Market, and Taizhou Electronics Market.

If Zhejiang could be labeled a province with "Three Nos", Taizhou was a microcosm of Zhejiang's predicament, housing still fewer natural resources than the provincial average. Moreover, Taizhou is one of the most inaccessible places in Zhejiang in terms of transportation, and with a per-capita land area of just one-sixth of the already low provincial average. Central investment in Taizhou during the planned economy had also been weak even by Zhejiang standards. Where the province was famous for its "Three Nos", Taizhou received the dubious adage of "Three Lines" (Zhang et al. 2005): the "war frontline" between mainland China and Taiwan, which endured sporadic military hostilities throughout the 1950s and 1960s; the "firing line" of the Cultural Revolution, in which local armed skirmishes flared into one of the most severe uprisings in communist China, and the "short line", a reference to receiving minimal state investment. From 1950 to 1978, central investment in Taizhou totaled less than half a billion yuan, the lowest of all regions of Zhejiang (Shi et al. 2004).

Amid this position of almost universal disadvantage, Taizhou's industrial culture and focus on machinery was born out of an interesting convergence of socio-historical circumstances. First, the market reform triggered mass emigration from Taizhou in search of new employment and business opportunities, which directly stimulated local demand for tools to equip migrants for the sectors they were entering. In the early 1980s, more than half a million people left Taizhou, about one-fifth of the local labor force (Chen and Dai 1996; Taizhou Statistics Bureau 2005). Shoe repairing was a popular service provided by many outgoing Taizhounese and Taizhou quickly became the largest production base in China for tools for this trade. From these beginnings, local machinery manufacturing later evolved to sewing machines, becoming one of Taizhou's major products.

Second, a unique recycling sector in Taizhou created a favorable environment for local machinery makers, both in terms of material supplies and learning to reverse engineer existing goods. Due to the disadvantaged local economy and lack of resources, during the immediate post-reform period a number of Taizhounese took up collecting, sorting, and reselling waste materials and scrap equipment from other parts of China. This business is described by Wei Xiajiu, party secretary of Taizhou from 1983 to 1987:

> At that time, we had a big business here. People from Luqiao and other places of Taizhou would collect waste metals from outside. The business was continuously growing. Abandoned machine tools, power generators, automobiles, tractors, as well as components from aircraft and tank were collected from everywhere in China. (Wei 2008)

According to Xiang Bingyan, his successor as party secretary from 1987 to 1990, this business developed beyond waste collection, to refurbishing and reselling complete machinery:

> I was truly surprised when I first saw this business in Luqiao. Inside and outside the town, on the streets and roads, near the agricultural lands, scrap machines were everywhere. Some stuff may weigh tens of tons. Where did they get these wastes? And how could they ship these things over? After all, roads to Taizhou were not in good condition and difficult to get through. These scraps were obsolete and oily. Waste plastics and rubber were piled like a mountain. Luqiao people worked very hard on these wastes. After disassembling, repairing, cleaning and painting, these machines became as though they were newly-made, and were then sold to small factories in villages. For

non-repairable ones, people broke them down and got the steel and other mineral materials, which were scarce and precious commodities, and could only be acquired through the national economic plan. This business was not only lucrative, but also supportive of the industrial development of Taizhou. (Xiang 2008)

Silver extraction offers a concrete example here. Once they became aware of the existence of silver in household and industrial waste, such as photographic negatives, used batteries, and various discarded machines, Taizhou people were able to extract tons of silver through recycling. More importantly, the process of disassembling and re-assembling machinery de-mystified the content and functions of these seemingly sophisticated industrial products for the local recyclers. Once familiar with how to strip down and rebuild machinery, local people were able to make their own. In consequence, the recycling sector brought about a distinct production and business culture centered on reverse engineering: local entrepreneurs would acquire mechanical products of interest, disassemble and figure out how they worked and were made, and begin producing similar equipment themselves.

Third, unique grassroots financial arrangements in Taizhou helped to resolve credit constraints encountered by private machinery manufacturers. As machinery is relatively capital intensive to produce, finance presented a major barrier for local entrepreneurs who were not only short of working capital, but effectively barred from the state-owned banking system. In response, Taizhounese entrepreneurs developed financial innovations that would enable them to raise the needed capital. The first of these was a stock-cooperation system, in which several individuals or households cooperated to establish and manage an enterprise together by mutual agreement. This type of cooperative system was heavily based on social networks. Through family and social ties, fathers and sons, brothers and sisters, relatives and friends worked together on new business opportunities. First emerging in Taizhou in the early 1980s, Enterprises with this institutional arrangement did not fit into established official categories. They were clearly not state-owned enterprises, nor could they be accurately termed "collective enterprises", owing to their relatively small size, nor "individual enterprises" because of the existence of multiple investors. Nonetheless, the stock-cooperation system was soon accepted by local officials and even considered a local institutional advantage in comparison with other nearby regions (Xiang 2008).

Active civil finance also helped to provide the necessary backing for local industrial projects. Extending from kinship and social ties, underground lending activities provided popular finance solutions in Taizhou. These commonly took the form of *hui*, or a kind of credit union. A typical *hui* consisted of one "head", or convener, who was the individual in need of capital, and several "feet", or lenders. Each of the "feet" would have a specific payment and withdrawal scheme, usually based on rolling dice. A *hui* could have many advanced varieties, some of which were profit driven and involved large funds. In addition to *hui*, another credit source was local private finance institutions. As state-owned banks tended to ignore private enterprises, especially small and medium-sized firms, private finance institutions expanded rapidly in Taizhou from the late 1980s. Taizhou is probably the only prefecture-level city in China that is home to as many as three private banks.

Cars: Passion of Li Shufu

Li Shufu, the father of Geely Auto, popularly known as the "automobile maniac" of Taizhou, is representative of the intense zeal of local private entrepreneurs, bent on producing machines as an escape route from the formerly minimal economic opportunities of the region. Even considered alongside the dynamic industrial culture of machinery in his hometown, Li's determination to produce cars appeared "audacious" and even "crazy" in the eyes of local people. After all, a modern car project generally involves massive investment and advanced technology, which would be extremely challenging to achieve through the efforts of a single entrepreneur. Seemingly contrary to all commonsense, Li Shufu appeared fearless in his gallant media statements about the project:

> How could it be so difficult to make a car? Isn't a car just a motorcycle plus two additional wheels? Isn't a car just two sofas with four wheels? (Author: this might be the most frequently quoted sentence from Li Shufu in the media.)
>
> Give me a chance to commit suicide! (Author: this is a well-known reply from Li when criticized that making cars might be tantamount to suicide.)
>
> What is innovation? Innovation should be something you insist on, even though all others disagree. If your decision is the same as others, how could it possibly be innovative at all? (Author: this statement depicts both Li's definition of innovation and his justification for the unusual decision to make cars.)

Li's enthusiasm for producing machines did not arise from nowhere, but was deeply rooted in the local industrial culture of Taizhou Manufacture. But his specific insistence on cars was unique among Taizhou entrepreneurs, and can only be understood from a personal perspective. Li's early business career provides important clues (Zheng 2007). In the 1980s, after graduating from the local high school aged 19, Li launched a photography business. Learning from the recycling sector, Li soon moved from taking photographs to extracting silver from film developer fluid, from which he made his first fortune. By the mid-1980s, Li had switched to making refrigerator components, in the process gaining sufficient experience that he decided to produce whole refrigerators. In the face of criticism and doubt from others around him, Li figured out how to make evaporators, the key component in refrigeration, made his first batch of complete refrigerators, and achieved good sales by the end of the 1980s. It was precisely during this period that the central government started to place controls on refrigerator production, expelling unapproved manufacturers from the market. Although there were ways of avoiding this regulation, such as seeking an alliance with a state-certified enterprise, Li saw no future in this pursuit and immediately gave up his refrigerator business without a fight. His career bounced back to impressive success in the early 1990s making magnalium sheet, a popular decorative constructional material. After accumulating a large investment fund, Li returned to machinery, this time choosing to make motorcycles, branded Geely. Within about 5 years his motorcycles were selling well and exported to more than 20 countries.

Like many Taizhou businessmen, Li's business interests always circled around a specific field: machinery. His entrepreneurial spirit was extraordinarily strong, regardless of criticisms and negative comments by family, peers, and others; his defiant attitude in the face of disparagement was the harbinger of his insistence on making cars. Moreover, Li's previous failure in refrigerator production may have strengthened his determination to make cars. Lessons learned in the refrigerator business may have led Li to be more flexible in the car project, and not to give up so easily.

However determined Li was to make cars, at the time this still ran counter to central government policy for the automobile sector, which prevented small producers from starting up in the belief that newcomers would have a negative effect on the already over-crowded industrial structure. Geely would have to fight hard for an official permit, while the same regulations made setting up a joint venture impossible; thus Li, like Chery Auto, had to achieve self-sufficient technological upgrading. Geely was even more disadvantaged than Chery in having no local government backing and thus no credit in the eyes of central administrators. So the political resources to persuade the central authorities to accept the project at a later stage, which was critical to the survival of Chery Auto, were not available to the maverick "automobile maniac" Li Shufu.

As for the extent that Geely might also count on the local governments of Zhejiang, unfortunately for Li, the project was met with a nonchalant attitude among local officials. Despite the general positivity surrounding the emerging private economy, the regulatory environment of the time made it impossible for local officials to support Li Shufu's project. Unlike the interventionist local government in Anhui, Zhejiang officials tended not to directly intervene in the market, but to leave business operations to private entrepreneurs; and due to the high risks involved in the car project, local officials were not at all optimistic about Li's prospects.

Geely Auto, Born Without Permit

Confronting national industrial policy while simultaneously receiving little local government support, the birth of Geely Auto occurred amid by far the most disadvantaged circumstances of the four cases in this study. Geely had no endorsement from the central government such as was enjoyed by relative giants FAW and SAIC, and no local official backing to see it through, as was the case for Chery Auto. Li and his colleagues had to tread water to survive the discriminatory policy environment, and the company struggled to achieve technological upgrading with no foreign partners to help introduce modern equipment and R&D.

Surviving Regulatory Discrimination

As a private car manufacturer, Geely Auto faced at least three major challenges to its initial survival under the existing government regulations: central administrators' aversion to giving new market entrants a production permit; access to funds to begin and sustain automobile production; and market opportunity for the final auto products. These interconnected issues threatened the survival of Geely: without legal status certified by a production permit, it would be difficult to acquire bank loans or other investments; the shortage of capital would drive bottlenecks in manufacturing activities and thus thwart any application for a production permit; and if Geely cars could not be sold in the market, the permit and investments would be meaningless anyway. None of these issues were easy for Li Shufu and his colleagues to tackle, yet equally, none deterred them.

In the four years from 1997 to 2001, Li deployed various tactics to persuade officials to approve his "illegal" car-making plan. His first strategy for getting a foot in the door of the auto industry, learned during his previous refrigerator and motorcycle endeavors, was "shell-borrowing"—attempting to affiliate Geely with other enterprises already certified by the state in order to gain use of their permit. In 1997, Li discovered an auto plant operated by a local jail in Sichuan province. The factory was meant to provide a vocational training program for prisoners, and was certainly far from a normal enterprise. However, Li could not afford to be picky about who he approached for a production permit, and was prepared to sacrifice a 30% share to persuade the factory leaders to enter the deal. This project was the first step in Li's car-making career, though it was less than satisfactory:

> Except for the merit of cheap labor (as the factory workers were prisoners), there were many problems in the jail factory. First, in a jail, it was not easy to get in and out. To make it worse, it was hard to match institutions between a private enterprise and a jail. They (the prison officials) did not contribute a penny (to this project); and in the prison, every tiny issue had to be reported upwards for directions. (Meng 2003: 96–97; content in brackets added by author)

After the initial trial in the jail factory, shell-borrowing was consistently deployed by Geely to deal with permit trouble. In 1997 when Li took steps to leave the jail factory and set up a car production base in Taizhou, his own motorcycle business was successfully used as a shell to acquire land in Linhai, a county-level city under the administration of Taizhou. With the permit and land acquired from the shells, Geely commenced and quickly expanded automobile production. The first model, branded Haoqing ("ambition" in Chinese) was brought out in 1998. The next year, in addition to the initial Linhai production base, Li built a second plant in Ningbo city, which produced the second Geely model, Merrie ("happy days" in Chinese) in the year 2000.

Despite these seemingly successful early steps, a serious problem soon arose. Production permits were only valid in the location where they were originally issued, and approval of both the Linhai and Ningbo plants risked being revoked. Getting a new permit was once again imperative. This time, Li switched strategies to one that might be called "crying for support": appealing for sympathy from the general public. He publicized the discrimination Geely had been treated to as a private enterprise, and sought official attention by publicly reiterating complaints for media consumption and dissemination:

> Please allow private enterprises to bravely explore, and allow us to have dreams of making cars! In comparison with official investments on cars of several billions, we will not ask for one cent from the government, and will not borrow one cent from the government. All outcomes will be on us and there is no risk on the side of government. Please give us a chance to fail!

Although the degree to which Li's public complaints affected the official mindset cannot be accurately assessed, it was quite evident that public opinion was beginning to exert pressure on national policymakers. Influenced by the media, the public started to show great sympathy towards this new carmaker, a mood echoed by scholars advocating the development of a private economy. In response, the central government eventually began to take this auto "activist" seriously. Four investigation teams were sent in succession by the central government to visit Geely from July to September in the year 2000. The visitors included significant policymakers such as Zeng Peiyan, director of the National Planning Committee. In 2001, Geely Auto was finally approved as a legitimate carmaker. Geely's struggles for public support contributed to the broader transformation of the Chinese automobile industry since the late 1990s.

Raising sufficient funds was another persistent challenge for Li. Despite being a successful businessman, Li's personal wealth was far from sufficient to finance a car project. It was hard for Geely as a private corporation to seek financial assistance from banks, which were mostly run by the state and issued loans preferentially to state-owned enterprises. Regardless of its identity as a private corporation, banks would not provide loans to Geely Auto due to its illegal status. Neither were local private finance institutions easily persuaded. Rejected by formal financing channels, Li sought alternatives. Firstly, he successfully gained assistance from the local civic financing network. Although the detailed investment structure in Geely's early years has not been made public, it was certain that Li Shufu was backed by a group of local private investors. Li developed a unique mode of working with some investors: rather than directly taking over the investments, Li mobilized these investors into component production to assist Geely. This was Geely's "boss project", as the media termed it (Yuan and Xie 2002), under which Geely invited interested local investors to act as "boss" to establish or jointly establish component enterprises to supply Geely Auto. Although these enterprises worked like sub-factories of Geely, the relationships between Geely and these investors were not based on a strict ownership structure or commercial contract, but on network-based informal negotiations. What mattered were social norms and trust among local businessmen. The "boss project" is a good illustration of how business behaviors were embedded in social networks. After Geely received its formal permit in 2001, production quickly expanded, and Li decided to raise further funds from the stock market, but in case his brand-new company did not attract enough attention from stock-market investors, Li again adopted a shell-borrowing strategy. Through a series of complex capital operations with Proper Glory, a public company in Hong Kong, as the shell, Li Shufu successfully raised funds from the Hong Kong stock market over about 3 years, from 2003 to 2005.

Finally, accessing sufficient market space was pivotal for the survival of Geely Auto in its infancy. Recall the situation of Chery: in the absence of a production permit in its early stages, local state actors in Anhui had no option but to create a market for Chery cars through incentives aimed at the local taxi business. This was not a feasible solution for Geely, however, as it was impossible for the local government of Zhejiang to intervene in the market on the part of a private carmaker. In fact, early Geely products could not even be sold as cars—the shell production permit "borrowed" by Li from the Sichuan jail factory came into the bus category. In this situation, the initial production run of Geely cars could not appear in any formal markets. Even when Geely later acquired a legitimate car-making permit of its own in 2001, market prospects were little more promising, as since the late 1990s, the Chinese car market had become crowded with competing foreign models introduced by joint ventures.

Li's response was to fight for a market by producing the cheapest car in China. Geely's first model, Haoqing, was sold at the exceptionally low price of about US$ 6000, occupying the cheapest price point in the market. After 2001, when the central government began to release price controls on cars, Geely immediately initiated price wars against its domestic competitors. In the face of increasingly numerous domestic carmakers, Geely tirelessly fought to maintain its position of making the cheapest cars in China. How could Geely produce cars at such extraordinarily low prices? In this regard, Geely shared some similarities with Chery. For example, Geely adopted the same "rolling development" principle practiced by Chery in its construction of manufacturing infrastructure and acquisition of equipment, saving significant investment by building infrastructure to rolling requirements. Also analogous to Chery, Geely's R&D was independent, so no extra charges to foreign corporations were incurred, as compared to products developed through the joint ventures. However, unlike Chery and many other carmakers in China, Geely's supply system offered unique advantages to further lower the final product price. A group of component manufacturers consistently supplied products at exclusively low prices to Geely, even without any ownership connections. This price distortion was mainly due to their trust-based expectations of long-term cooperation, as discussed above for the "boss project":

> Many component suppliers for Geely Auto were the previous suppliers of Geely's motorcycles. They had been transformed in order to fit the strategic development of Geely. By now, these suppliers, about 200 enterprises, were mainly concentrated in Taizhou, Ningbo and Wenzhou. Geely Auto controlled over a half of them, which is a primary reason for Geely to control the production cost well. (Li Shufu in Zhang 2004)

Besides these socially and historically connected suppliers, the low price of Geely cars was also supported by Zhejiang's thriving auto-parts sector. Since the market reform, a well-developed private auto-component cluster had emerged in Zhejiang, with private suppliers usually maintaining the low prices needed to compete against state-owned enterprises. It was thus relatively easy for Geely Auto to identify cheap components for use in its cars. On the back of this aggressive pricing, Geely Auto achieved and held onto a good position in the domestic car market in the early 2000s.

In dealing with these three major developmental barriers, Geely has actually progressed in the broader transformation of the Chinese automobile industry since the late 1990s. As previously discussed, China's automobile sector policies combined two seemingly incompatible rationales during the reform: the continuation of Mao-era economic nationalism on the one hand; and new ideas of bringing in foreign corporations on the other. The combination of these logics is reflected in China's first automobile industry policy issued in 1994 (see State Council of China 1994). There were three aspects to this policy: to facilitate import substitution through high tariffs; to upgrade obsolete technologies through joint ventures with foreign companies; and to concentrate the fragmented sector through entry regulations. These three aspects have been explored in the previous chapters on FAW, SAIC and Chery respectively. The policy aimed to provide a coherent policy package through which domestic automakers were expected to become self-reliant in R&D. Joint ventures were to work as the core mechanism to facilitate technological learning and upgrading, whereas the trade regulation and entry controls were supposed to create a positive business environment for both international and domestic actors. Chinese policymakers were all too aware of the potential risk that joint ventures could be controlled by foreign automakers, potentially inhibiting learning and upgrading among domestic firms. The central authority therefore set a precautionary 50% upper limit on foreign shares in any joint venture, prohibited foreign automakers from engaging in more than two joint ventures, and pushed for vigorous schedules within joint ventures to increase domestic content.

However, as the Chinese automobile market started to reach maturity by the end of the 1990s, the negative face of introducing joint ventures with foreign automakers emerged to much public criticism. Up to the mid-1990s, domestic demand for cars was mostly from institutional (i.e. primarily state) consumers, and the market was particularly distorted by high tariffs, state pricing controls, and the few oligarchical automakers that dominated supply. Foreign trade barriers, inflated, officially set prices, and little market competition presented opportunities for high profit margins to any foreign automakers engaging in joint-venture projects in China. By the late 1990s, nevertheless, the market had fundamentally shifted towards burgeoning private demand for passenger cars. Market competition intensified, official price regulations loosened, and China's entry to the World Trade Organization (WTO) in 2001 finally ended most previous restrictions on foreign trade. In response to this new business environment, global automakers started to take the Chinese market more seriously. From the late 1990s, foreign automakers starting and/or expanding their investments fought increasingly hard for larger market shares in China. In consequence, the rapidly enlarging Chinese automobile market was soon saturated with foreign-brand models rolled out under joint ventures with a few major international automakers, as described in the introduction of this book. This situation ran counter to the state's original policy aims for the sector, and incurred serious public backlash. Doubts were rising as to whether China's automobile policy was on the right track. A typical comment was that China's automobile sector was "large but not strong", because domestic automakers remained weak in self-reliant R&D capacity, in spite of increasing output overall.

The joint-venture strategy that had been so important to developing domestic production capability began to be called into question. For the public, it appeared ironic that independent automakers such as Chery and Geely performed more aggressively in developing national cars than officially endorsed automakers with the privilege of establishing joint ventures.

The fact that Chery and Geely had long been ignored and denied by national policies exacerbated the sense of this strange paradox in China's car industry, and generated great public sympathy toward firms that had been discriminated against. At the same time, huge political pressure began to bear down on the established joint ventures. The joint-venture strategy was criticized as a problematic arrangement, in which domestic automakers were diverted away from the greater national economic interest and towards the easy access to technology and profit-making opportunities offered by cooperation with foreign partners. Simultaneously, old-school management approaches towards the industrial structure of the auto sector also touched public nerves. The pursuit of industrial concentration by prioritizing a few key automakers was accused of creating oligopoly conditions and impeding normal market competition. Lastly, import substitution was blamed as an outdated policy measure, prompting rent-seeking to take advantage of domestic price distortions. Not only did these criticisms align with the fact that import regulations were no longer viable after China's WTO entry, but they appealed to the promotion of automobile exports by drawing on China's advantageously cheap labor.

A range of actors were behind criticisms leveled at Chinese automobile sector policy during this period. Former bureaucrats in the sector were most vocal in calling for a policy shift. With ingrained attitudes in favor of self-reliant R&D capacity, these retired officials, including He Guangyuan, previous head of the Ministry of Machinery, and Gen Shaojie, former director of FAW, were infuriated by the predominance of foreign models in the Chinese market, and strongly felt that the industry had betrayed its original direction. Due to their significant status, the voices of these officials swiftly reached central policymakers. Scholars also contributed to this discussion. Lu and Feng's notable policy report (2005), for example, was a timely and important study, revealing the unfair position of national car manufacturers such as Chery Auto in relation to joint ventures with larger enterprises. This report came out during major political conferences in China, generating strong sentiments against the approach of the large domestic automakers. The variety of responses by central ministries to this new policy discourse depended on their specific interests. The Ministry of Science and Technology, for instance, was active in sponsoring ideas in favor of technological self-reliance, and in fact had funded Lu and Feng's study mentioned above. The Ministry of Commerce, on the other hand, was less keen on the self-reliance issue, and more concerned with export promotion.

Automakers themselves exhibited a clear division in this debate. As described above, those who felt mistreated, such as Geely, could not pass up this chance to "cry for support", making their voices heard by the normally oblivious ears of established policy. Li Shufu was one of the most active participants in numerous policy discussions, appealing for the removal of policy barriers faced by private corporations in the car sector. On the other side, large domestic automakers holding joint ventures, who were now coming in for criticism, chose to defend their stance in the early stages of the debate. They called on the public to show patience, arguing that the scale of automobile production in China was still too small to afford self-reliant R&D, effort and investment in which could easily turn out to be wasted. Under growing political pressure, however, the large enterprises eventually gave in, promising to develop national brands in the near future.

This discourse was not unidirectional, however, and in the face of predominant criticism over the slow progress of self-reliant R&D under joint ventures, other experts warned against economic nationalism, sparking further debate. Perhaps the most famous instance is the so-called "He–Long debate" in 2005, which was a public discussion between He Guangyuan, the former head of the Ministry of Machinery, and Long Yongtu, China's chief representative in the WTO entry negotiations. He Guangyuan was an ardent advocate for self-reliant R&D and national-brand cars, whereas Long Yongtu insisted that the Chinese automobile industry should not overemphasize independence in the era of globalization. By cooperating with advanced foreign automakers, Long Yongtu proposed, China could participate in the international labor division of the global automobile industry, exporting automobile products to world markets.

The debate appeared never to reach a conclusion, although the new industrial policy issued in 2004 indicated a clear shift of emphasis within this important industrial sector (see State Council of China 2004). First, the Chinese government showed an increased tendency to promote self-reliant R&D and national brands. The 2004 industrial policy encourages domestic automakers to develop R&D capability, products with independent intellectual property rights, and national brands. Second, foreign trade policy has moved from import substitution to export promotion, by entitling "export bases" and "export enterprises" to government support, including Chery Auto as an export champion. Third, strict policy-led governance has been replaced by more market-oriented mechanisms. Past policy instruments such as investment approval procedures and directory management have been abandoned. Rather than designating certain automakers as key, the government has acknowledged the fundamental role of market competition in structuring the automobile sector. This policy transformation has improved conditions for firms such as Geely to prosper in the market, while also initiating broader new trends in the Chinese auto sector, as will be further discussed in the conclusion.

How Geely Made National-Brand Cars

> To build joint-ventures is like getting addicted to opium. The one who holds the brand and core technologies controls the game.
>
> We should make efforts to export Chinese cars to the world, rather than watch cars from all over the world getting into China.

These two oft-quoted statements by Li Shufu express his nationalistic stance on the car sector. Not only did Li strongly attack the joint-venture strategy advocated by national policymakers, he also dreamed of selling national cars to the international market. These claims are strongly reminiscent of similar calls issued by Chery director Zhan Xialai. It was no coincidence that the two entrepreneurs had the same ideals. For both Zhan Xialai of the state-owned Chery and Li Shufu of the privately run Geely, a negative attitude to joint ventures and a determination to be technologically self-sufficient stemmed from the same institutional logic: both had been excluded by national car-sector policy, and thus had no possibility of entering joint ventures. From a practical perspective, they had to be on their own.

As with Chery, Geely Auto's initial strategy was one of integrative imitation—copying an existing model and utilizing domestically available components as much as possible. Within Geely, this strategy was referred to as *miao hong*, a method of Chinese calligraphy training in which a template is followed to write a Chinese character. Unlike Chery, however, which carefully selected an ordinary model as its first template, Geely's approach was more aggressive: Li chose a luxurious Mercedes-Benz as his first template. The imitation project began in earnest in 1997, when Geely was still struggling to secure a production permit. By then, Geely's engineering team comprised three staff from Li's motorcycle factory. With no blueprints, specification data, or equipment, except for some drawing boards and a few hammers, the work was purely exploratory. Simply put, the team took apart one Benz and one Hongqi (the FAW car model), and combined their components into a new design, with bodywork hammered by hand from glass-reinforced plastic (Mei and Feng 2005). The next year, another model was put together in a similar manner, but this time Li's engineers made additional progress by developing a complete inventory of components to be used.

After these two bold experiments, in a rare departure from his usual character, Li decided to be more realistic and downgrade the target vehicle to a Xiali, a cheap but popular domestic model that was the main product of Tianjin Auto in the 1980s and 1990s. Tianjin Auto was one of the "Small Three" state-approved carmakers in China, once cooperating with Daihatsu and later being incorporated into the collaboration between FAW and Toyota. Just as Chery had imitated the Jetta from FAW, Geely began to comprehensively imitate the Xiali of Tianjin Auto. As a result, Geely's first Haoqing model borrowed extensively from the Xiali, including the chassis, engine, and many components. The second model, Merrie, and the third, Ulion, were also developed from the Xiali platform. In particular, all of these models were fitted with the same engine, the Toyota 8A, which had been developed for Xiali cars by Tianjin Auto.

As when Chery tried to follow an imitative R&D strategy with the Jetta, Geely also soon found itself hit by formal actions from the Xiali's manufacturer. Tianjin Auto, as the original producer of the Xiali and major victim of Geely's approach, was exasperated by the growth of the Geely models. From the perspective of Tianjin Auto, it was clearly ridiculous for Geely to take advantage of the entire supply system for the Xiali while also directly competing with the Xiali in the same market. In response, Tianjin Auto increased the price of the Toyota 8A engine which had been substantially used by Geely Auto in 2000; and later prohibited its engine factory and components suppliers from doing any business with Geely (Zhu 2006). In consequence, Geely had to search for alternative suppliers for components, and more importantly accelerate its own engine R&D. But a fresh round of actions against Geely soon followed. At the end of 2002, Geely was sued by Toyota over similarities between the logo of the Geely Merrie and that of Toyota, which it was felt could lead consumers to incorrectly associate the two firms. The legal conflict last for almost a year, with Geely refusing to compromise when faced with high compensation requests from Toyota.

Geely's early technological development, as well as the similar experience of Chery discussed in Chapter 4, underscores the value of technological imitation for Chinese independent carmakers, even in the face of formal actions that may result. For carmakers excluded by industrial policy and thus unable to acquire technology via joint ventures, there were precious few options for the first step in R&D other than imitation, a short-cut for inexperienced automakers to quickly start production and gain a foothold in the market. Although imitations often entailed potential infringement of intellectual property rights, there were also loopholes. Within the transitioning market environment of China in the 1980s and 1990s, the ambiguity of laws and policies, the cost of pursuing lawsuits, and concern not to offend the general public often constrained the victims, and helped illegal automakers to progress safely in the end.

Geely began to wind down its technological imitation phase by 2001 after receiving the production permit. In this new round of technological upgrading, Geely formalized procedures for whole-car R&D. In its previous imitations of the Xiali, Geely's R&D had entailed a reverse-engineering process with hand-drawing tools. In the new phase, Geely began to conduct forward engineering using digital models. Their engineers started to follow modern R&D procedures and standards: the Mybo model developed in 2002 represented the first complete forward-engineering attempt, and Free Cruise, a sedan model developed in 2005, marked Geely's complete accomplishment of this important R&D transformation (Mei and Feng 2005). In the meantime, Geely products began to evolve from targeting lower- to higher-end consumers. Geely's early models, such as the Haoqing, Merrie and Ulion were all low-end products, but from the Free Cruise onwards, Geely had broken into a more advanced market. Soon after the Free Cruise, Geely developed the Kingkong in 2006 and Vision in 2007. These two models plus the Free Cruise were known as "the new three" (as opposed to "the old three" of Haoqing, Merrie and Ulion), all of which were mid-level cars with more advanced technologies and higher price tags. By 2007, "the new three" had replaced "the old three" in the market, occupying over 60% of Geely's total sales (Fig. 5.2). This indicated a gradual transition in Geely's sales structure away from the original strategy of competing for the minimum price.

Geely had made important progress in component technology as well since 2001. Tianjin Auto's suspension of engine supplies acted as a catalyst, driving Geely to produce core auto parts such as engines and transmission independently. Based on Toyota engines, Geely developed its own engines to use for its cars. Soon afterwards, Geely independently developed two generations of engines. According to a domestic professional evaluation in 2007, three of Geely's engines were listed among the top Chinese automobile engines., They also achieved major breakthroughs in transmission technology. Automatic transmission had been a problem for Chinese carmakers since the 1980s, due to its complex technological requirements, with all automatic transmissions having to rely on direct imports or foreign technology transfers via joint ventures. In 2002, Geely Auto initiated R&D in this field. After three years, the project successfully produced China's first independently developed automatic transmission, which was then fitted to Geely's Free Cruiser model.

These achievements were based on a range of strategies. First, like Chery, Geely outsourced critical R&D to specialized international firms, enabling rapid technological upgrading of its cars. The search for external technological assistance began as early as 2002 with two technology providers in Italy and Korea, followed by invitations to other corporations to join the firm's R&D initiatives, whether domestic or international. Geely sought to use these activities to improve its own R&D team, capturing learning opportunities, as had Chery Auto, and often sending its own technicians to participate.

Also comparable to Chery, Geely continually sought to recruit experts in technology and management from all possible sources. Under Li Shufu, all four associate directors of Geely Auto were external recruits: Xu Gang, a former general accountant at Zhejiang Local Taxation Bureau; Nan Yang, a

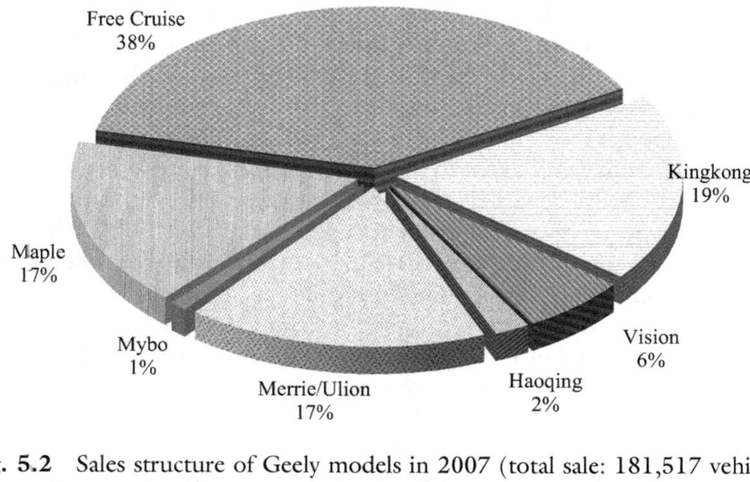

Fig. 5.2 Sales structure of Geely models in 2007 (total sale: 181,517 vehicles)
Note: The Maple model was developed by another private carmaker, which was incorporated to Geely Auto in 2002
Source: Geely Auto 2008

former general manager at Shanghai-VW; Zhang Zhe, a former general accountant with Macro Group; and Zhan Wanjin, a former general manager of business at FAW-VW (Su 2007). The profiles of these Geely associate directors are reminiscent of Yin Tongyao, also a management expert from FAW-VW who later moved to Chery Auto. Geely also carried out strategic headhunting of important technicians, beginning with an engineering team "invited" from Tianjin Auto to assist with imitation of the Xiali. This is again similar to Chery's approach, in which a group of technicians from DMC was recruited to assist in bringing its early models to production. Key Geely technicians who had been invited from outside positions included Pan Yanlong, former general engineer of the Nanjing Fiat joint venture. Joining Geely to take charge of the firm's R&D Institute, Pan became the principle founder of Geely's R&D system. Xu Binkuan, previously the general engineer at Tianjin Gear Factory and director of a national research team on automatic transmission, was hired to develop automatic transmission. Cui Xiaoli, former director of the technology department of Tianjin Toyota Motor Corporation, led the research team at Geely which successfully developed the VVT engine. Hua Fulin, previously a chassis expert at FAW, became the technician in charge of Geely chasses. Finally, Guo Konghui, a renowned Fellow of the China Academy of Engineering and China's top expert in automobile engineering, joined Geely as a senior technology consultant.

Although the developmental progress of Geely has been marked by business strategies designed to combat a discriminatory policy environment, the Zhejiang local government was not a totally negative or irrelevant agent with regard to the company's growth. It is true that local government actors were initially unwilling to get involved in the Geely project. No financial assistance was available through state agencies, and even land for Geely infrastructure was only approved under the name of Li's already successful motorcycle business. However, once Geely acquired a legitimate production permit in 2001 and began to grow in a promising way, the local government showed support for the project, albeit retaining a hands-off approach to its specific operations. By then, top political leaders in Zhejiang began to publicly acknowledge the value of Geely Auto and demonstrate their intention of offering necessary assistance, which proved a helpful shift in mindset for Geely. For example, the land approval in Ningbo for Geely's second production base proved much smoother than before. Interestingly, local government assistance was sometimes more personal than institutional, unlike the Chery "businessman with red hat" case, some local officials resigning their government positions to join the carmaker.

In recent years, Geely Auto has maintained a promising development trajectory. In terms of market performance, Geely started with sales of 24,000 vehicles in 2001, rising to about 400,000 in 2010 and over 500,000 in 2015 to rank among the top ten manufacturers in China. New models have been rolled out to expand the Geely fleet, including the Gleagle, Emgrand, and Englon, while Li's production bases have expanded far beyond Zhejiang Province to major regional centers including Shanghai, Jinan, and Chengdu. Geely Auto has also recently made important moves towards upgrading to high-end models. In 2010, Geely purchased 100% equity in the globally renowned Volvo Car Corporation of Sweden, an event symbolic of the rapid rise of the Chinese automobile sector. In 2013, Geely took over Manganese Bronze Holdings, the producer of London's iconic black cabs, aiming to transform production to an electric version. Four R&D centers have been set up by Geely Auto in Shanghai; Barcelona, Spain; Gothenburg, Sweden; and California, United States. In addition, like Chery, Geely has been actively exploring overseas markets, exporting about 10,000 vehicles per year in 2012 and 2013 to countries such as Russia, Ukraine, Saudi Arabia, Iran, and Egypt. Geely Auto has survived against significant institutional and economic odds to become one of the most successful grassroots carmakers in China.

Case Summary

Geely Auto, the most successful grassroots national brand, is a representative of the last production model, in which a private carmaker developed and manufactured cars independently. This chapter has examined the origins of the prosperous private economy of Geely Auto's home province of Zhejiang, whose former national positioning under the planned economy had a somewhat skewed association with its subsequent market transition. Featuring a rugged landscape, minimal resources, and a defensive border with Taiwan, the economy of Zhejiang was long overlooked in the pre-reform era, yet the resulting shallow legacy cultivated a surprisingly active local private economy during the market reform, largely out of pure necessity for local entrepreneurs to make a living. The relentless car manufacturing ambitions of Li Shufu, founder of Geely Auto, led to a unique local industrial culture dominated by machine making in Taizhou city. Geely's autonomous approach to car production, like Chery Auto's, was initially in serious violation of central regulations aimed at industrial concentration, but in the subsequent sociopolitical process Geely Auto fought for legitimacy as a private carmaker and developed its manufacturing capabilities independently of state support or foreign partners. In the face of a discriminatory administrative environment, as a private company Geely encountered even greater difficulties than Chery, which was at least backed by proactive local officials. Various strategies were therefore employed by Geely to access production permits and realize technological upgrading, including outsourcing R&D capacity and overseas purchases. As well as depicting Geely's progress in establishing itself through this production model, the chapter points to how the endeavors of Geely have contributed to recent shifts in the governance of the Chinese automobile sector releasing entry controls and promoting national brands, as evidenced by the new industrial policy launched in 2004.

Bibliography

Chen, H. (1997). Good Bamboo Root Comes from Bitter. In Zhejiang Historical Archive Committee (Ed.), *Small Commodity and Large Market: Memory of the Builders of the Yiwu Small Commodity Market* (pp. 177–189). Hangzhou: Zhejiang People's Press. (in Chinese).

Chen, J., & Dai, X. (1996). *Taizhou, from a Follower to a Pioneer*. Beijing: Economy Daily Press. (in Chinese).

Ding, L. (2007). *The 2007 Report of Regional Comprehensive Competitiveness in Guangdong*. Guangzhou: Social Science Academy of Guangdong. (in Chinese).

Fang, M. (2000). *Institutional Change and Developmental Track of Zhejiang*. Hangzhou: Zhejiang People's Press. (in Chinese).

Gao, S. (1987). Regional Problem in the New Stage of National Economic Growth. *Management World, 4*, 26–40. (in Chinese).

Geely Auto. (2008). *Corporate Presentation*. Hangzhou: Geely Auto. (in Chinese).

Huang, Y. (2008, October 20). Zhejiang Province: A Free-Market Success Story. *Business Week*.

Lu, F., & Feng, K. (2005). *The Policy Choice to Develop Chinese Automobile Industry with Independent Intellectual Property Rights*. Beijing: Peking University Press. (in Chinese).

Mei, Y., & Feng, K. (2005). Making Cars in Geely: A Report of Self-Reliant Innovation of Geely. *China Soft Science, 11*, 1–10. (in Chinese).

Meng, H. (2003). *Made by Private Enterprises: The Road of Geely and the Situation of Chinese Private Enterprises*. Guangzhou: Guangdong Tourism Press. (in Chinese).

National Statistics Bureau. (1982). *China Statistics Yearbook 1981*. Beijing: China Statistics Press. (in Chinese).

Pan, M. (1997). The Statement of the Fortune-Making Secretary. In Zhejiang Historical Archive Committee (Ed.), *Small Commodity and Large Market:*

Memory of the Builders of the Yiwu Small Commodity Market (pp. 101–108). Hangzhou: Zhejiang People's Press. (in Chinese).

Shi, J., Wang, W., & Qian, T. (2004). *Private Economy and Institutional Innovation: A Study on Taizhou Phenomenon*. Hangzhou: Zhejiang University Press. (in Chinese).

State Council of China. (1994). *The Industry Policy of the Automobile Industry*. State Council of China. (in Chinese).

State Council of China. (2004). *The Industry Policy of the Automobile Industry*. State Council of China. (in Chinese).

Su, Z. (2007). The Competitiveness of Geely. *China Logistics, 7*, 44–53. (in Chinese).

Taizhou Statistics Bureau. (2005). *Taizhou Statistics Yearbook 2005*. Beijing: China Statistics Press. (in Chinese).

Wei, X. (2008). *Taizhou People Created the Stock Cooperation System*. http://www.tzdsw.net.cn/newsinfo.asp?ID=477. Accessed 10 June 2017. (in Chinese).

Xiang, B. (2008). *The Right Developmental Way Should Fit Local Reality*. http://www.tzdsw.net.cn/newsinfo.asp?ID=477. Accessed 10 June 2017. (in Chinese).

Yuan, W., & Xie, D. (2002). The Fragile Geely. *South Reviews, 3*, 53–55. (in Chinese).

Zhang, W. (2004). *Build Geely as the Chinese Version of Hyundai in Korea*. http://auto.sohu.com/ 20041120/n223092912.shtml. Accessed 10 June 2017. (in Chinese).

Zhang, R., & Yang, Y. (2006). Zhejiang Merchants: Growth Background, Group Characters, and Developmental Trend. *Business Economics and Administration, 6*, 3–7. (in Chinese).

Zhang, M., Xie, S., & Guo, Y. (2005). *Introduction of New Taizhou Merchants*. Changchun: Jilin University Press. (in Chinese).

Zhejiang Statistics Bureau. (2006). *Zhejiang Statistics Yearbook 2006*. Beijing: China Statistics Press. (in Chinese).

Zhejiang Statistics Bureau. (2007). *Private Economy Become Critical Engine of Socio-economic Development of Zhejiang*. http://www.stats.gov.cn/ztjc/ztfx/dfxx/200712/t20071225_33874.html. Accessed 10 June 2017. (in Chinese).

Zheng, Z. (2007). *Li Shufu, Automobile "Maniac": Decoding the First Private Carmaker*. Beijing: China Citic Press. (in Chinese).

Zhu, J. (1988). Short of Resource but Rich in Economy: Some Thoughts on Strategy of Economic Development in Zhejiang. *Exploration, 2*, 16–33. (in Chinese).

Zhu, Q. (2006). The Vigor of Geely. *Ceocio China, 9*, 51–59. (in Chinese).

CHAPTER 6

Conclusion and Discussion

Car manufacturing is symbolic of modern industry. Production and sales of passenger cars in China are nowadays the largest in the world, and tracking the trends that brought about this outcome encapsulates China's rise in the global economy. This book has sought to shed light on the far-reaching transformation of China's automobile industry since the 1980s, and in particular the variation among Chinese carmakers in terms of how production has been organized. The book develops an institutional explanation, highlighting that national economic positioning, a historical arrangement of the Chinese central government under the planned economy, in large part shaped the post-reform development of various local automobile projects, and ultimately brought about the diversified production models seen in China's car sector today. Within this path-dependent process, local political structures and the configuration of local developmental ideas were the two major routes by which the institutional legacy of national economic positioning exerted effects. Empirically, the book applies this framework to explore four representative cases: First Auto Works (FAW), Shanghai Automotive Industry (Group) Corporation (SAIC), Chery Auto, and Geely Auto. Detailed accounts of each case are presented in the four previous chapters, revealing how car manufacturing was influenced and catalyzed by pre-reform national economic positioning.

Q. Feng, *Variety of Development*,
https://doi.org/10.1007/978-981-10-5912-4_6

Summary of Findings

When investigating economic activity, sociologists often focus on nations or sectors, rather than specific individual enterprises. However, for automobile production, which is often centralized under a few major entities, researchers have sometimes been able to reveal the dynamics of the broader sector through a small number of key cases (e.g. Harwit 1995; Thun 2006). Using four representative cases, this study has examined the transformation of the Chinese automobile industry in the past three decades, which has been at best partially investigated in previous literature.

By any measure, the recent development of China's automobile sector is remarkable. From the end of the 1990s, it took China about ten years to overtake all the former champions of global car manufacturing and become the largest automobile producer in the world. Questions naturally follow as to who is behind the explosive rise of automotive production in China, and how they have achieved it? There are no simple answers, however. Behind the sheer pace of the rise of China's car industry lay a legion of assemblers in significantly higher numbers than those seen in other countries. As mentioned in previous chapters, the total number of Chinese automotive assemblers had reached over 100 by the end of the 1980s, only declining very recently to about 60, yet still much larger than the international norm: by the 1980s, worldwide independent assemblers had already dwindled to about 30 due to market competition and consolidation. Even within the several major automakers at the top of the pile in China in relation to the numerous peripheral assemblers, there remains great diversity in how automobile production has been organized.

This book systematically explores the diversity of production arrangements by analyzing ownership structure and technological strategy, two principle dimensions of automobile production for a developing country such as China. It is observed that, in terms of ownership structure, manufacturers either belonged to the central government, local governments or private capital; regarding technological strategy, some firms pursued the introduction of contemporary foreign models through joint ventures, while others took an independent path to manufacture national-brand cars.Four distinct production models are readily identifiable from these varying configurations.

To decode the variety of production arrangements among Chinese manufacturers matters not only because of the present magnitude and status of China's automobile sector, but also for its theoretical relevance in understanding institutional variations within a given industrial sector under conditions of economic reform and globalization. The production models observed in the Chinese automobile sector appear somewhat puzzling to a number of established literatures in that they exhibit quite contradictory patterns: the assemblers who have been most aggressive in developing national-branded cars in China are generally small and receive minimal government support, whereas large state-owned automakers have mostly committed to the production of foreign cars introduced through transnational joint ventures. Current literature on market transition and globalization has at best provided limited explanations for the divergence of Chinese automakers. Set against the backdrop of economic reform and opening up to foreign direct investment, the Chinese car sector may reasonably be assumed to be dominated by private enterprises and foreign products. But the persistent presence of state-owned automobile enterprises and the rise of national-brand cars in China plunges these otherwise plausible scenarios into disarray.

Intriguingly, this pattern also appears to challenge a significant body of literature on technological upgrading of enterprises. Whether one focuses on firm size as highlighted in the Schumpeter Hypothesis, international affiliation in terms of global value chains, or presence of state support based on the theory of developmental states, each perspective encounters difficulties in understanding China's automobile sector. In developing national-brand cars, the firms favored by these theories—large with abundant resources for innovation, partnered with foreign automakers to enjoy learning opportunities, and supported/endorsed by the state to access policy advantage for upgrading—performed weakly in practice by comparison with those that lacked these conditions. These "ideal" firms arguably achieved greater levels of technological sophistication by producing mature models introduced by their global partners. However, the advanced capabilities of these firms stopped short of some smaller national carmakers, who from sheer necessity made faster progress in Research and Development (R&D) for whole-car systems, a key marker of independent manufacturing capacity; practiced systematic technical innovation; and perhaps more importantly, showed strong entrepreneurship. These patterns enabled valid explanations of the variety of production arrangements among Chinese carmakers, contrary to theoretical predictions, since variables such as firm size, foreign affiliation, and government support could have been controlled for.

The book provides an institutional explanation, highlighting the path dependency of local automobile sectors in China extending from their national positioning in the former planned economy. From this perspective, the varying production models of today's Chinese automobile sector did not result from the exercise of free choice by firms in the new market environment, but are traceable to the national institutional arrangements established in the early years of planned economy. National economic positioning, which created the scaffold against which China's economy structured and molded the functions of local economies, has had a lasting effect through the gradualist market transition of China. For local automobile sectors in the reform era, this institutional legacy shaped post-reform power structures and the configuration of developmental ideas among the major stakeholders, underpinning the socio-political process surrounding the establishment and development of car manufacturing projects, and finally contributed to the specific production model of each entity.

Specifically, this book proposes that for regions with core positions on the previous national economic map, central–local affiliations tended to be strong during the market transition, so that for local car initiatives, the political structure either featured a leading local state with central sponsorship or the direct leadership of the central government, and the local developmental ideas of these related actors were often closely coordinated. As a result, local projects to manufacture passenger cars would often gain significant hierarchical support, or even become a flagship project of the central government. In contrast, previously marginalized regions rarely enjoyed such advantages, so that when car projects were established, local governments or private enterprises had to take the lead, often without higher acknowledgment or endorsement. This would spiral into an administrative conflict when the central government perceived these spontaneous projects as operating against state interests in national industrial concentration. In these instances, local car projects fought for survival under (or often outside) the strict surveillance of the central administration.

This book has analyzed production models among Chinese carmakers using four representative cases. Overall empirical findings are summarized in Table 6.1, in which the three major institutional factors of pre-reform national economic positioning, local political structure, and configuration of local developmental ideas in the reform, as well as the subsequent socio-political process of establishing car production, are compared across the empirical cases. The ownership structure and technological strategy of the four carmakers—the two defining aspects of the production models—are interpreted by examining these institutional factors alongside China's transformation in the market reform.

Table 6.1 Summary of the four empirical cases

Production model and empirical case	*Pre-reform national economic positioning*	*Local political structure during reform*	*Configuration of local developmental ideas during reform*	*Socio-political process*
Model 1: First Auto Works (FAW)	Northeast: the national heavy-industry base	**Central government**: The leading power in critical issues of the local economy **Local governments**: Coordinator of central administrative policy **FAW**: Follower of the central administration	**Central government**: Import substitution was necessary and joint venture was acceptable **Local governments**: The car sector was important to the local economy **FAW**: It was a duty to follow national industry policy	The central government transformed FAW as a central enterprise, and encouraged FAW to establish joint ventures with foreign partners. The local governments and FAW closely followed the central guidance
Model 2: Shanghai Automotive Industry (Group) Corporation (SAIC)	Shanghai: the national economic center	**Municipal government**: The active builder of the local economy with extraordinary political power **Central government**: Supporter of the municipal government **SAIC**: Follower of the municipal government	**Municipal government**: Establishing automobile joint ventures was key to reinvigorating the local economy **Central government**: Development of Shanghai should be supported, thus so should joint ventures **SAIC**: The direction of the municipal government should be followed	The municipal government of Shanghai made great efforts to establish its own car enterprise and launched joint venture s with foreign companies. These initiatives were supported by the central government, and implemented by SAIC
Model 3: Chery Auto	Anhui: a resource-contributing province to serve other regions (such as Shanghai)	**Local governments:** The dominant actors in charge of local economic development **Central government:** Provider of autonomy through fiscal decentralization **Chery Auto**: Follower of the local governments	**Local governments:** The car sector was a breakthrough for the long underdeveloped local economy **Central government:** Unapproved car projects would disturb national industrial concentration **Chery Auto**: The ideas of the local governments should be fulfilled	Under surveillance (and initial restriction) by the central administration, Chery Auto was covertly established by the local governments of Anhui, later remaining technologically independent, without foreign partners
Model 4: Geely Auto	Zhejiang: a marginalized region with minimal central government investment	**Private enterprises**: The most active force in the local economy **Local governments:** Laissez-faire administrators of the local private economy **Central government**: Provider of autonomy through fiscal decentralization	**Private enterprises**: Machinery manufacturing was a good business to pursue, as was car-making **Central government**: Unapproved car projects would disturb national industrial concentration **Local governments:** State should not intervene in private business operations	Geely Auto as a private carmaker struggled for legal status from the central administration, with limited local government assistance. Geely Auto remained technologically independent without joint ventures with foreign corporations

As can been seen from Table 6.1, the production models of the four cases historically result from the different economic functions assigned to their geographical locations by the central government of China in the pre-reform era, through which national planners shaped the architecture of the command economy. The comparison of these four cases reveals a typology of various economic positions in China's earlier planned economy, the structure of which was far from even but decidedly hierarchical. Due to its long-established status in the colonial era, Shanghai, the location of SAIC, was designated the national economic center, concentrating economic activities requiring sophisticated and advanced technology, and contributed the largest share of national fiscal income under the planned economy. Similarly, Changchun, Jilin Province, the birthplace of FAW, is located in China's northeast which had been assigned the role of the national base for heavy industry. Though the northeast is still not comparable to Shanghai in terms of its economic position, it is hard to overestimate the importance of the northeast for China's industrial sectors. Considering heavy industry as the engine of wider modernization, throughout the planned-economy period, the central administration invested intensively in this region, nurturing China's largest cluster of state-owned enterprises. In contrast to Shanghai and Changchun, the home locations of the other two cases were low priorities, and largely ignored on the national economic map. Chery Auto's home province of Anhui had been assigned a supply role in the national economy, contributing materials and resources to other regions, especially Shanghai, connected by the Yangtze River. If Anhui was overlooked in many ways in this role, the prospects of Zhejiang province, the origin location of Geely Auto, were especially dim. As a province with negligible natural resource endowments and in close proximity to a potential conflict front with Taiwan, Zhejiang received the least central attention in terms of economic planning and investments in industry, so that the mechanisms of the planned economy so influential elsewhere were sometimes barely functional in this region.

As China's economic reform began to take hold and car production became a focal area of investment, the contrasting national economic positions of the early years steered the four cases along distinct developmental pathways. Firstly, the legacy of national positioning shaped post-reform power dynamics among local and central actors related to car manufacturing. Due to characteristic economic duties that had long been institutionalized in the planned economy, local political structures took on various forms in the market transition, in which prior advantages still mattered.

For the northeast, the previous national base for heavy industry to which the central government had committed huge, long-term investments, tight state control persisted during the reform, and was the leading force in the post-reform development of FAW. At the other end of the scale, Zhejiang Province, long ignored by central planners and with little investment in the planned-economy era, never had such historical hierarchical ties. The weak legacy of planned-economy institutions instead promoted a vibrant private economy during the reform, with numerous energetic entrepreneurs arising as the most active force in the local political economy. Shanghai and Anhui provide another pair of comparison in this regard. Once the central government loosened controls over their respective local economies via decentralization during the reform, the local governments of Shanghai and Anhui both became masters of their economic destinies. However, the historically important status assigned to Shanghai gave local officials leverage to gain substantial support from central planners, which was not possible for an ordinary province such as Anhui. The local political structures of Shanghai and Anhui therefore varied according to the degree of backing from the central authority, which later led to differing fates in their plans to harness the car industry to stimulate local economic development.

Pre-reform economic positioning also affected local patterns of developmental ideas relating to car projects, which generated specific rationales and strategies among related actors. The main ideas of the central government about the automobile industry during the early stages of reform were embodied in its 1994 industrial policy, which worked as a national guideline with the key developmental ideas of promotion of import substitution, introduction of foreign technology via joint venture, and concentration of the industrial structure. However, each of these ideas was received differently at the local level. Automakers from previously centralized positions in the national economy, such as FAW and SAIC, were endorsed by the central government as key players, and thus were in line with the national industrial policy, assigned as the privileged few in the national team to initiate large-scale passenger car production for import substitution. As such, they were the first to be permitted and encouraged to launch joint venture with foreign companies, considered a necessary national strategy to acquire the necessary technology, and a turning point in opening a vital industrial sector to foreign involvement in China. For these two cases, the configuration of local developmental ideas was primarily collaborative and without friction. However, officially legitimatizing certain automakers such as FAW and SAIC also implied the exclusion of "outsiders". For this reason, Chery and Geely, despite strong local motivations to establish car projects, were initially unable to gain permission from the central administration to do so. The configurations of local developmental ideas in these cases were therefore filled with tensions, in which spontaneously established automakers went against the national policy, to boost the local economy and/or seek access to a lucrative business opportunity.

The ways that specific ownership structure and technological strategy were adopted in the four Chinese carmakers are demonstrated through local socio-political processes, in which related actors—central and local governments, enterprises, and businessmen—interacted in the local car projects, with different political powers and developmental ideas. In the case of FAW, the story is mainly one of top-down manipulation by the central government in the market reform. By comparison, for Geely Auto, it was grassroots private entrepreneurship that drove progress under the radar of the central administration. Consequently, while FAW continued as a major centrally managed enterprise, implementing the national strategy of introducing technology from joint ventures with foreign partners, Geely Auto later became a leading private automaker, producing national-brand cars with independent R&D. On the other hand, a local developmental state was the common theme for the socio-political processes surrounding the rise of SAIC and Chery Auto. For the former, the process was relatively smooth, with the historically prestigious municipal government eventually receiving endorsement by the central administration, even for its then unprecedented proposal to launch joint ventures in the automobile sector. However, for the latter, the local government received no such special treatment, thus had to strategically weather a harsh regulatory environment for its car project, which breached the national automotive policy to concentrate the industry structure. In consequence, SAIC and Chery Auto, though both owned by the local governments, adopted contrasting technological strategies: one in partnership with foreign automakers via joint ventures, and the other forced to manufacture national cars independently.

Research Limitations

This section reflects on some of the key limitations of this study. First, while it is accurate to consider the four cases examined as representative of the four production models, they should not together be considered fully representative of the Chinese automobile sector—this would be not only overly simplistic but risk excluding other manufacturers and assemblers from the picture. That said, the firms (shown in Fig. 1.2) not selected for each production model usually shared a similar developmental trajectory to the selected case. For example, among the three enterprises under Model 1, the case of Dongfeng Motor Corporation (DMC) closely follows FAW. Established by the central government in the late 1960s, DMC was intended as a back-up factory for FAW, which was to maintain automobile production in the hinterland in the event of war. After the reform, like FAW, DMC was transformed into a centrally managed enterprise, and granted permission to initiate joint ventures with foreign corporations, first with PSA Peugeot Citroën (PSA) in 1992, then Kia in 2002, and Nissan and Honda in 2003. Changan Auto, another enterprise in Model 1, also pursued a broadly similar track, though with a more specialized military background. As one of the earliest military enterprises of China, the Changan plant was relocated to the Southwest of China during the second Sino-Japanese war (1931–1945) and maintained as a major military factory under direct central administration in the planned economy. When the reform started, following calls from the central government to convert to civil production, Changan Auto established joint ventures with Suzuki in 1993, and later with Ford in 2001, Mazda in 2007, and PSA in 2011. Changan Auto thus had a similar story to FAW, except that the path dependency of Changan Auto was less about the previous national economic positioning, and more associated with its unique military affiliation.

Similarly, two additional automakers under Model 2, Beijing Automobile Industry Holding Corporation (BAIC) and Guangzhou Automobile Industry Group Corporation (GAIG), could be regarded as lower-performing versions of SAIC. Thun (2006) differentiates these assemblers from SAIC as failed cases, as the municipal governments of both Beijing and Guangzhou, though playing leading roles in promoting the local automobile sector, suffered from fragmented institutions and incentives, and thus performed poorly in nurturing the local supply network in the 1990s. In consequence, their local automobile sectors were long managed in an unstable and fractured manner. Not until 2000 was BAIC finally officially established, and the first joint venture of Guangzhou, Guangzhou-PSA, finally went bankrupt in 1998 after 13 years of operation. Despite weak state capacity and limited market performance as compared with Shanghai, both Beijing and Guangzhou were firmly sponsored by the central government. Their automakers were consistently counted as the national "Small Three" throughout the 1980s and 1990s, and permitted to engage in joint ventures with foreign companies. As early as 1984, Beijing Jeep was established in cooperation with American Motors Corporation as the first joint venture in the Chinese automobile industry, though the main product was jeeps as opposed to passenger cars. Guangzhou-PSA was started only a year later, in 1985, the same year that SAIC-VW was established. And there was no policy barrier for them to expand the list of foreign partners in the following years: joint ventures were launched with Hyundai in 2002 and with Daimler Chrysler in 2003. And in Guangzhou, following the demise of Guangzhou-PSA, GAIG soon received approval from the central government to cooperate with Honda in 1998, and with Toyota in 2004. As regards the active local state with support of the central government, BAIC and GAIG are similar to Shanghai in nature. However, the explanation of historical economic positioning in the case of Shanghai may only apply to Beijing, which similarly enjoys political leverage due to its supreme status as the national capital, but not for Guangzhou, which gained special favor from the central government only after the reform commenced.

Brilliance Auto, co-listed with Chery Auto under Model 3, is something of an outlier. Shenyang city in Liaoning Province, the hometown of Brilliance Auto, is a major industrial city of the Northeast. In the late 1980s, the main local assembler, Shenyang Automobile Factory, started to produce light buses with technological assistance from Toyota. A private entrepreneur, Yang Rong, took over this enterprise in the 1990s, and started work to develop a national-brand car branded Brilliance in the year 2000. As with Chery, the Brilliance passenger car was not initially acknowledged by the central administration. However, while still seeking a production permit, Yang Rong was dismissed due to an ownership dispute, and the local government reacquired Brilliance Auto. The endeavors of local officials later obtained an official production permit in 2002, and a joint venture with BMW was set up in 2003. Despite these twists of fate, Brilliance cars continued production, becoming an important national model in the early 2000s. Although categorized under Model 3, Brilliance Auto thus mixes characteristics of Model 2 and Model 4: it appears similar to Geely Auto because of the involvement of a private entrepreneur and a well-established national brand, whereas it also resembles Model 2 in the decisive role of the local government.

In addition to Geely Auto, Model 4 includes BYD Auto, a significant private Chinese manufacturer with national-brand models. Both cases shared important commonalities. Similar to Li Shufu, the charismatic father of Geely, BYD was born out of the ambition and effort of a successful private entrepreneur, Wang Chuanfu. Wang's BYD (an abbreviation of "Build Your Dreams") started out making cellphone batteries in Shenzhen, neighboring Guangzhou. As BYD progressed to become a major global player in battery production by the early 2000s, Wang became determined to produce electric vehicles using BYD's advanced battery technology. This quickly attracted widespread criticism and doubt, similar to those cast over the "maniac" Li Shufu and Geely. Encountering the same problem as Geely over obtaining an official production permit, Wang purchased Qinchuan Auto in 2003, a small automaker in northwest of China, as his "shell" enterprise to gain legal status. BYD could then set about proving itself in the domestic automobile market. Its first model, the F3, achieved initial market success in 2005, while the plug-in hybrid model F3DM rolled off the assembly line in 2008, and the fully electric model e6 commenced production in 2010. Though resembling Geely, BYD is a latecomer with a unique focus on new alternate energy vehicles, thus receiving less policy discrimination. More importantly, like Guangzhou, Shenzhen as the hometown of BYD gained special attention from central planners after the reform. Although the pre-reform foundations of the local private economy in Shenzhen may not be as strong as in Zhejiang, as the experimental region of the central government for a market economy since the late 1970s, Shenzhen turned out to be a playground for China's private entrepreneurs, which offered the necessary environment for BYD to emerge. This latter aspect bears stronger resemblance to Geely's origins in Zhejiang.

China's automobile sector encompasses numerous relatively small-sized enterprises which are not investigated in this book. Some of the smaller firms have been incorporated into the larger automakers. For example, two of the more prominent small firms, Hafei Automobile Group and Jiangxi Changhe Automobile Corporation, were recently merged into Changan Auto. In addition, a few unlisted automakers have shown strong growth potential and thus deserve attention in future studies. Great Wall Motors is one such case. As a private automaker in Hebei province, Great Wall began manufacturing pick-up trucks in the 1990s, expanded to sport utility vehicles (SUVs) in the early 2000s, and recently made good progress in the production of sedan models. A similar product development trajectory can be seen in Anhui Jianghuai Automobile Co. (JAC), a local state-owned automaker in Anhui Province. This smaller provincial neighbor of Chery Auto transferred production from buses and light trucks in the 1990s to multi-purpose vehicles (MPVs) and SUVs in the early 2000s, and recently also began to manufacture sedans. These two cases suggest an interesting possible upgrading pathway in the smaller-scale Chinese automobile sector, in which manufacturers of commercial vehicles move to producing SUVs and/or MPVs, and finally become capable of developing sedan cars.

As noted in Chapter 1, to classify Chinese automakers through our four production models only captures the main features of production arrangements among these enterprises, as in practice, the ownership structure and technological strategy may be much more complicated than these ideal types. This is particularly true given the recent policy transformation reflected in the 2004 industrial policy under which the central government began to release market entrance controls, promote exports, and more importantly encourage national brands. In this sense, domestic automakers can be observed to show an increasing tendency to switch tracks in terms of the ways they have organized production up to now. In response to this official call for national cars, the "Big Three" started to move towards self-reliant brands. Besides continuously expanding production of the historical Hongqi model, for example, FAW introduced a new national car, Besturn, in 2006, which was developed in cooperation with Mazda Auto. In the same year, SAIC rolled out the Roewe, its first national model, albeit purchased from the British MG Rover Group, which went bankrupt in 2005. Later in 2007, SAIC added another model of MG in its fleet of national models. Though the output of FAW and SAIC in terms of national-brand cars is not yet comparable to their established foreign joint venture models, it seems that automakers under Model 1 and 2 have begun to progress along the long-expected trajectory towards self-reliant development envisaged by the policymakers.

Meanwhile, national carmakers under Models 3 and 4 have also developed tendencies to cooperate with foreign companies via joint ventures in a new environment free of state regulation. For example, in 2012, Chery Auto set up its first joint venture with Jaguar Land Rover Automotive PLC, a strategic move to promote its current status from low-end to high-end production of passenger cars. Interestingly, at about the same time, BYD Auto established a joint venture with the Daimler Group. Diverging from the old form of joint ventures, in which the Chinese firms were often on the technologically dependent side, this initiative is based on the mutual needs of BYD Auto and the Daimler Group to produce new alternate energy vehicles. That is, BYD aims to learn advanced car production technologies from Daimler, while Daimler also has a great interest in BYD's electric vehicle battery technology. In the face of these new trends, the production models of Chinese automobile manufacturers may continue to experience significant changes in future, probably evolving beyond what has been described in this book.

Theoretical Implications

The analyses of the diverse developmental trajectories of Chinese automakers presented in this book have several important theoretical implications. By examining a series of institutional factors, this study represents an attempt to integrate historical and sociological institutionalism as the two major schools of new institutionalism currently prevalent in the social sciences, and demonstrates the complementarities between them. In this sense, this study contributes to ongoing endeavors to reconcile different institutional theories. The theoretical framework of the book by and large follows the principles of historical institutionalism, which concentrates on the constraining effects of historically formed institutional contexts, and the resulting path dependency of future changes (see Thelen 1999; Campbell 2004). By unearthing the association between national economic positioning under China's command economy and the subsequent multiple local configurations of power and developmental ideas under the market reform, this study reveals how the post-reform regional automobile sectors in China were heavily shaped by the legacy of their previous national designations.

On the other hand, the sociological institutionalism perspective has also been applied to cognitive schemes and ideational discourse as constitutive of organizational behaviors (Meyer and Rowen 1977; DiMaggio and Powell 1983). In this regard, the way that developmental ideas stimulated and nurtured the local actions has been a consistent, central explanatory component of this book. Moreover, the mixture of historical and sociological institutionalism in this study is further reflected in the varying mechanisms of ideas as discussed across the empirical cases. Both the historical and sociological institutionalist schools acknowledge the role of ideas in action. According to Campbell (1998, 2002), specifying the types and mechanisms of ideas in institutional analysis can promote "cross-fertilization" between historical and sociological institutionalism. In this direction, it is noteworthy that local developmental ideas functioned through differing mechanisms across the cases under study. In the case of FAW, the ideas of the central government as embodied in national industrial policy functioned as a regulatory and directing force, thus more in alignment with historical institutionalism. In contrast, Geely Auto was more revealing of mechanisms of sociological institutionalism in that the initiative to produce passenger cars was cognitively nurtured by the unique local industrial culture. For the remaining two cases, SAIC and Chery Auto, constraining and constitutive mechanisms of ideas co-existed. That is, while the enterprises' respective home locations of Shanghai and Anhui Province were both under central policy regulation, local officials had undergone unique ideational discourses in policymaking, markedly different in consequence.

In considering the production arrangements that result from various local discourses, this book has adopted a constructivist perspective, which has implications for current organization studies. In the 1980s and 1990s, setting up a new car plant in China, no matter how "rational" this may appear nowadays, was fraught with risk and uncertainty. The book deconstructs such presumed rationality in each case, by investigating how carmaking initiatives originated out of the debates that engaged related actors in each local unique context. From this perspective, the supposed rationality of launching a local automobile initiative amounts to nothing more than shared meanings grounded in a process of social construction (Berger and Luckmann 1966). Such a constructivist view has been well developed in current organization studies. Institutional approaches to organization studies propose that "rationalized" organizational practice, rather than being derived from any efficiency-driven calculations, is mostly based on concerns of legitimacy (Meyer and Rowan 1977). In a so-called organizational field where "a community of organizations that partakes of a common meaning system and whose participants interact more frequently and fatefully with one another than with actors outside of the field" (Scott 1994: 207–8), organizations become isomorphic though coercion, norms, or imitation (DiMaggio and Powell 1983). Nevertheless, as has been the approach in this book, the organizational field, if applied in analysis, should be regionalized due to differing local discourses, and should also comprise both state and corporate actors rather than purely economic organizations. More importantly, these organizational fields may not necessarily be internally harmonious towards an isomorphic trend, but contentious when related to actors' perceived contrasting interests, as is visible in the development of the local car projects examined in this study.

Despite adopting a constructivist approach in explanation, this study does not assume that ideas themselves can gain political currency without agency. Agency has become something of a weakness in institutional theory, since scholars of organization studies have emphasized the structuration of institutional forces over recent decades, while sometimes ignoring the agency of actors. The resulting deterministic role of institutions over agency has drawn significant criticism. For example, as Schmidt (2008: 314) sharply commented, current institutionalism assumes "'unthinking' actors who are in an important sense not agents at all". In consequence, institutionalists often encounter difficulty in accommodating agency in analysis, especially to explain how institutions change. To build a theory of action for institutional analysis, DiMaggio (1988) introduced the concept of the institutional entrepreneur, arguing that "new institutions arise when organized actors with sufficient resources see in them an opportunity to realize interests that they value highly" (p. 14). Since then, continuous efforts have been made to develop and expand this notion to accommodate the structure–agency tension in institutional studies (Hardy and Maguire 2008; Weik 2011). In this study, agency is highlighted within the analytical framework. No matter how persuasive car project ideas appeared to be, what matters was "how specific actors carried certain ideas into the policy-making fray and used them effectively" (Campbell 2002: 29). In the four local car projects investigated, developmental ideas were generated and mobilized via agency, and political structures indeed functioned and manifested through agency, though agency was simultaneously bounded and guided by these institutional structures. This intricate relationship has been extensively captured in the socio-political process elucidated in each case, where actors strategically mobilized developmental ideas and political power in response to the new policy and market environment in the reform. This study hence provides an effort to address the paradox of embedded agency, which refers to the theoretical dilemma of observing institutionally shaped actors to understand change in institutions themselves (see Seo and Creed 2002; Battilana et al. 2009).

This book also provides detailed comparison to enrich current discussions about market transition in China. The multi-tracked organizational dynamics of Chinese automakers under the economic reform reject any simplified, linear pathway towards a market economy. However, regarding the mechanism and nature of the market transition, the book provides favorable evidence for both the pro-market and corporatist views on China's market transition, the former emphasizing market expansion as the dominating force while the latter argues for the resilient role of the state (Keister and Borelli 2012). On the one hand, the collapse of centralized economic commands over production, the corporatization of Chinese automakers, and the rise of a national automobile market all indicate the significant progress of marketization in the Chinese automobile sector, which is in agreement with pro-market predictions. In particular, the case of Geely Auto provides a solid illustration of the bottom-up scenario depicted by pro-market scholars, in which private entrepreneurs struggled through state regulation and finally changed the attitudes of central policymakers. This aligns well with the recent work of Nee and Opper (2012), which theorizes endogenous institutional changes in market transition based on an empirical investigation of private entrepreneurs in the Yangtze delta region of China. On the other hand, the other production models and development trajectories represented by FAW, SAIC, and Chery Auto suggest the continued importance of the state in the Chinese automobile sector. The case of FAW embodies the persistent control of the central government, whereas SAIC portrays a proactive municipal government. More interestingly, the case of Chery Auto intensively captures the tensions between different levels of government regarding a local car initiative. The latter three cases thus tend to be more in line with a corporatist perspective. The multiplicity of production models among Chinese automakers, as revealed in this study, therefore indicates great heterogeneity in China's market transition across different regions and sectors. Introducing a sub-national lens on China's market transition may thus push current debates within the pro-market and corporatist arguments to a new level.

The framework adopted in this study also matches well with the concept of "new regionalism" coined by Rithmire (2014). Using recent studies on comparative political economy in China, Rithmire (2014) reviewed a trend to consider the local political economy of China as an endogenously evolving system. Rithmire's major examples included but were not limited to Susan Whiting (2001) on the diversified models of rural economies in Shanghai, Jiangsu, and Zhejiang; William Hurst (2009) on regional differences in China regarding lay-offs in the state-owned sector; and Eric Thun (2006) on the varying local supply networks of the Chinese automobile industry, also reviewed in this book. According to Rithmire (2014), these scholars tend to take the central–local relationship in China as not deterministically rigid, and "instead of using local realities to shed light on how China's national political and economic system works, they are examining how *local*-level variables determine different *local* economic realities" (p. 170). This book aligns with such principles, by consistently concentrating on "local-level variables", such as configurations of power and ideas among the related actors, to explain variation of production arrangements among Chinese automobile manufacturers, which are in turn deeply rooted in pre-reform national economic positioning. The fact that China's old planned economy was ironically far from uniform, but a highly uneven system integrating various economic functions, may be a trigger for future studies of regional diversity in China.

Although the impacts of globalization on the Chinese automobile sector cannot be overstated, this study provides a negative case against the globalist view that the role of the state will decline with globalization. Present advanced globalization has been mostly driven by the worldwide diffusion of neoliberal ideology since the 1970s, epitomized by Washington Consensus policy prescriptions first exported to Latin American countries in the early 1980s (Williamson 1990), and later across East and Southeast Asia in the late 1990s (Beeson and Islam 2005). For critics of the neoliberal ideal of globalization, the Washington Consensus has failed to achieve the central goal of promoting economic growth (McCleery and De Paolis 2008). More importantly, against neoliberal predictions, there are persistent national variations in the organization of market economies (Hall and Soskice 2011), and even with neoliberal policies distributed throughout the world, outcomes have proven to be markedly dissimilar across different countries (Fourcade-Guorinchas and Babb 2002). This book is in agreement with this line of thought: that the ultimate dominance of globalization over domestic institutions appears an unlikely prophecy. In tandem with the entry and expansion of global automakers, as this book reveals, the central government of China maintained persistent and tight control over the automobile sector through direct ownership of key enterprises, strict regulation of industrial structure, and deliberate technological strategies within joint ventures, in order to keep the actions of foreign corporations in line with national interests. Thus, globalization did not function in this instance in a top-down manner that would supposedly optimize economic activity, but was substantially moderated by the central administration through industrial policies according to national priorities (see Chin 2010).

Over and above the interplay of globalization and domestic institutions, the analysis presented in this book is further complicated by the fact that China is concurrently experiencing ongoing market transition. To explain the origin of varying production models of Chinese automakers thus represents an attempt to shed light on how globalization and marketization have intertwined at the national, regional, and organizational levels. Marketization created opportunities for global automakers to modernize outdated domestic automobile production, whereas globalization facilitated the emergence of the market mechanism in the automobile sector. This study has highlighted how the trajectories of both globalization and marketization in China are historically embedded in the institutions of national economic positioning in the communist period. As detailed in the empirical chapters, Chinese automakers, simultaneously exposed to incoming international firms and the newly emerging domestic market environment, pursued diversified pathways of development based on their respective regional legacies inherited from the command economy. In this sense, region-specific, historically embedded institutions formed the critical architecture of China's transitional economy in globalization, aligning between international and domestic structural changes to drive localized trajectories of development. Investigating associated topics along this dimension may lead to a more exact picture of how economic organizations may evolve in macro institutional and structural changes, both in China and elsewhere.

Bibliography

Battilana, J., Leca, B., & Boxenbaum, E. (2009). How Actors Change Institutions: Towards a Theory of Institutional Entrepreneurship. *The Academy of Management Annals, 3*(1), 65–107.

Beeson, M., & Islam, I. (2005). Neoliberalism and East Asia: Resisting the Washington Consensus. *Journal of Development Studies, 41*(2), 1–22.

Berger, P. L., & Luckmann, T. (1966). *The Social Construction of Reality: A Treatise in the Sociology of Knowledge*. Harmondsworth: Penguin Books.

Campbell, J. L. (1998). Institutional Analysis and the Role of Ideas in Political Economy. *Theory & Society, 27*(3), 377–409.

Campbell, J. L. (2002). Ideas, Politics, and Public Policy. *Annual Review of Sociology, 28*, 21–38.

Campbell, J. L. (2004). *Institutional Change and Globalization*. Princeton: Princeton University Press.

Chin, G. T. (2010). *China's Automotive Modernization: The Party-State and Multinational Corporations*. New York: Palgrave Macmillan.

DiMaggio, P. (1988). Interest and Agency in Institutional Theory. In L. Zucker (Ed.), *Institutional Patterns and Organizations: Culture and Environment* (pp. 3–12). Cambridge, MA: Ballinger.

DiMaggio, P. J., & Powell, W. W. (1983). The Iron Cage Revisited: Institutional Isomorphism and Collective Rationality in Organizational Fields. *American Sociological Review, 48*(2), 147–160.

Fourcade-Guorinchas, M., & Babb, S. L. (2002). The Rebirth of the Liberal Creed: Paths to Neoliberalism in Four Countries. *American Journal of Sociology, 108*(3), 533–579.

Hall, P. A., & Sockice, D. (2011). *Varieties of Capitalism: The Institutional Foundations of Comparative Advantage*. New York: Oxford University Press.

Hardy, C., & Maguire, S. (2008). Institutional entrepreneurship. In R. Greenwood, C. Oliver, R. Suddaby, & K. Shalin-Anderson (Eds.), *Handbook of Organizational Institutionalism* (pp. 198–217). Thousand Oaks: Sage.

Harwit, E. (1995). *China's Automobile Industry: Policies, Problems and Prospects*. New York: M.E. Sharpe.

Hurst, W. (2009). *The Chinese Worker after Socialism*. Cambridge, UK: Cambridge University Press.

Keister, L. A., & Borelli, E. P. (2012). Market Transition: An Assessment of the State of the Field. *Sociological Perspectives, 55*(2), 267–294.

McCleery, R. K., & De Paolis, F. (2008). The Washington Consensus: A Post-Mortem. *Journal of Asian Economics, 19*(5–6), 438–446.

Meyer, J. W., & Rowan, B. (1977). Institutionalized Organization: Formal Structure as Myth and Ceremony. *American Journal of Sociology, 83*, 340–363.

Nee, V., & Opper, S. (2012). *Capitalism from Below: Markets and Institutional Change in China*. Cambridge, MA: Harvard University Press.

Rithmire, M. E. (2014). China's "New Regionalism": Subnational Analysis in Chinese Political Economy. *World Politics, 66*(1), 165–194.

Schmidt, V. A. (2008). Discursive Institutionalism: The Explanatory Power of Ideas and Discourse. *Annual Review of Political Science, 11*, 303–326.

Scott, W. R. (1994). Conceptualizing Organizational Fields: Linking Organization and Societal Systems. In H. Derlien, U. Gerhardt, & F. W. Scharpf (Eds.), *Systemrationalitat und Partialinteresse* (pp. 203–221). Baden-Baden: Nomos Verlagsgesellschaft.

Seo, M., & Creed, W. E. D. (2002). Institutional Contradictions, Praxis, and Institutional Change: A Dialectical Perspective. *Academy of Management Review, 27*(2), 222–247.

Thelen, K. (1999). The Historical Institutionalism in Comparative Politics. *Annual Review of Political Science, 2*, 369–404.

Thun, E. (2006). *Changing Lanes in China: Foreign Direct Investment, Local Governments, and Auto Sector Development*. New York: Cambridge University Press.

Weik, E. (2011). Institutional Entrepreneurship and Agency. *Journal for the Theory of Social Behaviour, 41*(4), 465–481.

Whiting, S. (2001). *Power and Wealth in Rural China: The Political Economy of Institutional Change*. New York: Cambridge University Press.

Williamson, J. (1990). *Latin American Adjustment: How Much Has Happened?* Washington, DC: Institute for International Economics.

Bibliography

Acs, Z. J., & Audretsch, D. (1987). Innovation in Large and Small Firms. *Economic Letters, 23*, 109–112.

Aghion, P., Bloom, N., Blundell, R., Griffith, R., & Howitt, P. (2005). Competition and Innovation: An Inverted-U Relationship. *Quarterly Journal of Economics, 120*(2), 701–728.

Amsden, A. (1989). *Asia's Next Giant: South Korea and Late Industrialization.* New York: Oxford University Press.

An, H., & Ren, Z. (2008). Reconsideration of China's Traditional Planned Economic System from the Perspective of Twice Division of Government Power in the Past. *Journal of Sichuan University (Social Science), 2*, 95–101. (in Chinese).

Aoki, M., Kim, H.-K., & Okuno-Fujiwara, M. (1998). *The Role of Government in East Asian Economic Development.* New York: Oxford University Press.

Archive of First Auto Works. (2004). *FAW Yearbook 2003.* Changchun: First Auto Works. (in Chinese).

Arthur, W. B. (1989). Competing Technologies, Increasing Returns, and Lock-In by Historical Events. *Economic Journal, 99*, 116–131.

Bair, J., & Gereffi, G. (2001). Local Cluster in Global Chains: The Causes and Consequences of Export Dynamics in Torreon's Blue Jean Industry. *World Development, 29*(11), 1885–1903.

Bao, X. (1992). Developing Characteristic Products Is an Important Way of Invigorating Machinery Industry. *Economic Management, 8*, 16–17. (in Chinese).

Battilana, J., Leca, B., & Boxenbaum, E. (2009). How Actors Change Institutions: Towards a Theory of Institutional Entrepreneurship. *The Academy of Management Annals, 3*(1), 65–107.

Q. Feng, *Variety of Development*,
https://doi.org/10.1007/978-981-10-5912-4

Beeson, M., & Islam, I. (2005). Neoliberalism and East Asia: Resisting the Washington Consensus. *Journal of Development Studies, 41*(2), 1–22.

Bennett, D. C., & Sharpe, K. E. (1985). *Transnational Corporations Versus the State: The Political Economy of the Mexico Auto Industry*. Princeton: Princeton University Press.

Benziger, V. (1998). Can China's Gradualist Reform Strategy Be Applied in Eastern Europe? *Journal of the Asia Pacific Economy, 3*(1), 38–60.

Beyer, J. (2010). The Same or Not the Same: On the Variety of Mechanisms of Path Dependence. *International Journal of Social Sciences, 5*(1), 1–11.

Biggart, W., & Guillén, M. (1999). Developing Difference: Social Organization and the Rise of the Auto Industries of South Korea, Taiwan, Spain and Argentina. *American Sociological Review, 64*(5), 722–747.

Blanchard, D. R., Krugman, P., Layard, R., & Summers, L. (1991). *Reform in Eastern Europe*. Cambridge, MA: MIT Press.

Buchanan, J. M., Tollison, R. D., & Tullock, G. (1980). *Towards a Theory of Rent-Seeking Society*. College Station: Texas A&M University Press.

Cai, L. (1992). *The Choice Toward Modernization: The General Principle of Economic Development of Shanghai*. Shanghai: Shanghai Far East Press. (in Chinese).

Campbell, J. L. (1998). Institutional Analysis and the Role of Ideas in Political Economy. *Theory & Society, 27*(3), 377–409.

Campbell, J. L. (2002). Ideas, Politics, and Public Policy. *Annual Review of Sociology, 28*, 21–38.

Campbell, J. L. (2004). *Institutional Change and Globalization*. Princeton: Princeton University Press.

Cao, L., Gu, G., & Liu, J. (1987). *The Study on the Ownership Structure of Production Means in Shanghai*. Shanghai: Shanghai Social Science Academy Press. (in Chinese).

Cardoso, F. H., & Faletto, E. (1979). *Dependency and Development in Latin America*. Berkeley: University of California Press.

Central Committee of the Communist Party of China. (1999). The Decision of the Central Committee of the Communist Party of China on the Reform and Development of the State-owned Enterprises. *Announcement at the Fourth Plenary Session of the 15th Central Committee of the CPC*. (in Chinese).

Cerny, P. G. (1994). The Dynamics of Financial Globalization: Technology, Market Structure and Policy Response. *Policy Sciences, 27*(4), 319–342.

Chandler, A. (1990). *Scale and Scope: The Dynamics of Industrial Capitalism*. Cambridge, MA: Belknap Press.

Chang, H.-J. (2003). *Globalization, Economic Development and the Role of the State*. New York: Zed Books.

Chen, H. (1997). Good Bamboo Root Comes from Bitter. In Zhejiang Historical Archive Committee (Ed.), *Small Commodity and Large Market: Memory of the Builders of the Yiwu Small Commodity Market* (pp. 177–189). Hangzhou: Zhejiang People's Press. (in Chinese).

Chen, J. (1998). The Analyses and Suggestions for First Auto Works to Introduce the Lean Production. In Historical Record Office of First Auto Works (Ed.), *Studying in Japan* (pp. 198–241). Changchun: First Auto Works. (in Chinese).

Chen, Y. (2008). Why Chery Helds Enthusiasm for Innovation. *China New Time, 6*, 86–89. (in Chinese).

Chen, J., & Dai, X. (1996). *Taizhou, from a Follower to a Pioneer*. Beijing: Economy Daily Press. (in Chinese).

Cheng, Y. (1995). Start a New Age for the Chinese Car Industry: Inspirations Form the Santana Project. In The Propaganda Center of SAIC and Shanghai Auto News (Ed.), *Rising Up by the Huangpu River: A Report from the Shanghai Automotive Industry* (pp. 1–11). Shanghai: Shanghai Wenyi Press. (in Chinese).

Chin, G. T. (2010). *China's Automotive Modernization: The Party-State and Multinational Corporations*. New York: Palgrave Macmillan.

China Auto News. (2005). *Chinese Automobile Industry Full Record, 1949 to 2004*. Beijing: Chinese Auto News. (in Chinese).

China Automotive Technology Research Center. (1999). *The China Automotive Industry Yearbook 1999*. Tianjin: China Automotive Yearbook Press. (in Chinese).

Chinese Central Television. (2005). *Chery Contributes to the Competitiveness of the Chinese Automobile Industry*. http://www.cctv.com/program/dialogue/20050309/101611.shtml. Accessed 10 June 2017. (in Chinese).

Chinese Central Television. (2006). *Oriental Son: Yin Tongyao, Director of Chery Auto*. http://news.sina.com.cn/s/2006-06-29/212710289386.shtml. Accessed 10 June 2017. (in Chinese).

Coe, N. M., Dicken, P., & Hess, M. (2008). Global Production Networks: Realizing the Potential. *Journal of Economic Geography, 8*(3), 271–295.

Cohen, J. N., & Centeno, M. A. (2006). Neoliberalism and Patterns of Economic Performance, 1980–2000. *The Annals of the American Academy of Political and Social Science, 606*, 32–67.

Cohen, W. M., & Levin, R. C. (1989). Empirical Studies of Innovation and Market Structure. In R. Schmalensee & R. D. Willig (Eds.), *Handbook of Industrial Organization* (Vol. II). Amsterdam: North-Holland.

Committee of Shanghai Automobile Industry Record. (1999). *Shanghai Automotive Industry Record*. Shanghai: Shanghai Social Science Academy Press. (in Chinese).

Committee of Shanghai Automotive Industry History. (1991). *Shanghai Automotive Industry History: 1901 to 1990*. Shanghai: Shanghai People's Press. (in Chinese).

Committee of Shanghai Fiscal and Taxation Record. (1995). *Shanghai Fiscal and Taxation Record*. Shanghai: Shanghai Social Science Academy Press. (in Chinese).

Committee of Shanghai Local Record. (2005). *Shanghai Local Record*. Shanghai: Shanghai Social Science Academy Press/Shanghai People's Press. (in Chinese).

Cox, R. W. (1987). *Production, Power and World Order*. New York: Cambridge University Press.

David, P. A. (1985). Clio and the Economics of QWERTY. *American Economic Review, 75*(2), 332–337.

Dent, C. M. (2003). Taiwan's Foreign Economic Policy: The 'Liberalization Plus' Approach of an Evolving Developmental State. *Modern Asian Studies, 37*, 461–483.

Dicken, P. (2003). *Global Shift: Reshaping the Global Economic Map in the 21th Century*. New York: The Guilford Press.

Dicken, P., Kelly, P., Olds, K., & Yeung, H. W. (2001). Chains and Networks, Territories and Scales: Towards an Analytical Framework for the Global Economy. *Global Networks, 1*(2), 89–112.

DiMaggio, P. (1988). Interest and Agency in Institutional Theory. In L. Zucker (Ed.), *Institutional Patterns and Organizations: Culture and Environment* (pp. 3–12). Cambridge, MA: Ballinger.

DiMaggio, P. J., & Powell, W. W. (1983). The Iron Cage Revisited: Institutional Isomorphism and Collective Rationality in Organizational Fields. *American Sociological Review, 48*(2), 147–160.

Ding, L. (2007). *The 2007 Report of Regional Comprehensive Competitiveness in Guangdong*. Guangzhou: Social Science Academy of Guangdong. (in Chinese).

Dobbin, F. (1993). The Social Construction of the Great Depression: Industrial Policy During the 1930s in the United States, Britain, and France. *Theory and Society, 22*(1), 1–56.

Dong, Z. (2004). Examination of Old Industrial Base Invigoration Based on Characteristics of Infrastructure Investment of the Northeast in the Latter Half of the Twentieth Century. *Study of History of the Chinese Communist Party, 5*, 50–57. (in Chinese).

dos Santos, T. (1970). The Structure of Dependence. *American Economic Review, 60*(2), 231–236.

Economic and Trade Commission of Anhui Province. (1999). The Great Achievement of Industrial Development. In the Anhui Government (Ed.), *Fifty Years of Anhui* (pp. 329–334). Beijing: China Statistics Press. (in Chinese).

Editorial Committee of Anhui Yearbook. (2000). *Anhui Yearbook 1999*. Hefei: Anhui Yearbook Press. (in Chinese).

Estrin, S. (2002). Competition and Corporate Governance in Transition. *Journal of Economic Perspectives, 16*(1), 101–124.

Evans, P. (1997). The Eclipse of the State? Reflection on Stateness in an Era of Globalization. *World Politics, 50*(1), 62–87.

Fan, L. (2003). Businessman with Red Hat in Anhui. *Decision and Information, 10*, 19–20. (in Chinese).

Fang, M. (2000). *Institutional Change and Developmental Track of Zhejiang*. Hangzhou: Zhejiang People's Press. (in Chinese).

Feng, K., Yin, T., & Wang, Y. (2007). The Innovation Mode of Chery. *China Soft Science, 3*, 76–84. (in Chinese).

First Auto Works. (2003a). *First Auto Works Fifth Years Memorabilia, 1953 to 2003*. Changchun: First Auto Works. (in Chinese).
First Auto Works. (2003b). *The Path of First Auto Works in Fifty Years, 1953 to 2003*. Changchun: First Auto Works. (in Chinese).
Fligstein, N. (1996). Market as Politics: A Political-Culture Approach to Market Institutions. *American Sociological Review, 61*(4), 656–673.
Fligstein, N. (2003). *The Architecture of Markets: An Economic Sociology of Twenty-First-Century Capitalist Societies*. Princeton: Princeton University Press.
Fourcade-Guorinchas, M., & Babb, S. L. (2002). The Rebirth of the Liberal Creed: Paths to Neoliberalism in Four Countries. *American Journal of Sociology, 108*(3), 533–579.
Fu, Y. (2004). The Strategy and Construction of the Central Government in Northeast China in 1950s. *History Study of the Chinese Communist Party, 5*, 58–64. (in Chinese).
Fu, G., & Zhang, R. (2006). Self-Reliant Innovation: Strategy Under the Weakened Industry Barriers. *Contemporary Finance & Economics, 9*, 79–83. (in Chinese).
Gallagher, K. S. (2006). *China Shifts Gears: Automakers, Oil, Pollution, and Development*. Cambridge, MA: MIT Press.
Gao, Z. (1986). Exploring the Industrial Structure of Anhui. *Theory Research, 2*, 16–22. (in Chinese).
Gao, S. (1987). Regional Problem in the New Stage of National Economic Growth. *Management World, 4*, 26–40. (in Chinese).
Ge, B. (2007). Details of Negotiations on Santana. *Auto Business Review, 7*, 118–125. (in Chinese).
Ge, S., & Xia, S. (1999). The Fifty Year Transformation and Characteristics of the Industrial Structure in Anhui. *Jianghuai Forum, 6*, 30–33. (In Chinese).
Geely Auto. (2008). *Corporate Presentation*. Hangzhou: Geely Auto. (in Chinese).
Gen, S. (1993). Push Reform, Enlarge Opening, Catch Opportunity, and Make Accomplishment. In Historical Record Office of First Auto Works (Ed.), *First Auto Works Yearbook 1993* (pp. 29–40). Changchun: Jilin Science and Technology Press. (in Chinese).
Gereffi, G. (1999). International Trade and Industrial Upgrading in Apparel Commodity Chain. *Journal of International Economics, 48*(1), 37–70.
Gereffi, G. (2015). Global Value Chains, Development and Emerging Economies. *Inclusive and Sustainable Industrial Development Working Paper Series (WP 18)*. United Nations Industrial Development Organization.
Gereffi, G., & Kaplinsky, R. (2001). The Value of Value Chains: Spreading the Gains from Globalisation. *Special Issue of the IDS Bulletin 32*. The Institute of Development Studies.
Gereffi, G., & Korzeniewicz, M. (1994). *Commodity Chains and Global Capitalism*. Westport: Praeger.

Gereffi, G., & Memedovic, O. (2003). *The Global Apparel Value Chain: What Prospects for Upgrading by Developing Countries* (Sectoral Studies Series). Vienna: United Nations Industrial Development Organization.

Gereffi, G., Humphrey, J., & Sturgeon, T. (2005). The Governance of Global Value Chains. *Review of International Political Economy, 12*(1), 78–104.

Giuliani, E., Rabellotti, R., & Pietrobelli, C. (2005). Upgrading in Global Value Chains: Lessons from Latin American Clusters. *World Development, 33*(4), 549–573.

Goldstein, J., & Keohane, R. (1993). *Ideas and Foreign Policy: Beliefs, Institutions and Political Changes.* New York: Cornell University Press.

Gore, C. (2000). The Rise and Fall of the Washington Consensus as a Paradigm for Developing Countries. *World Development, 28*(5), 789–804.

Green, A. E. (1992). South Korea's Automobile Industry: Development and Prospects. *Asia Survey, 32*(5), 411–428.

Guillén, M. F. (2001). Is Globalization Civilizing, Destructive or Feeble? A Critique of Five Key Debates in the Social Science Literature. *Annual Review of Sociology, 27*, 235–260.

Hall, P. (1989). *The Political Power of Economic Ideas: Keynesianism Aross Nations.* Princeton: Princeton University Press.

Hall, P., & Taylor, R. (1996). Political Science and the Three New Institutionalisms. *Political Studies, 44*(5), 936–957.

Hall, P. A., & Soskice, D. (2011). *Varieties of Capitalism: The Institutional Foundations of Comparative Advantage.* New York: Oxford University Press.

Hamilton, G., & Biggart, N. W. (1988). Market, Culture, and Authority: A Comparative Analysis of Management and Organization in the Far East. *The American Journal of Sociology, 94*(Supplement), S52–S94.

Hardy, C., & Maguire, S. (2008). Institutional entrepreneurship. In R. Greenwood, C. Oliver, R. Suddaby, & K. Shalin-Anderson (Eds.), *Handbook of Organizational Institutionalism* (pp. 198–217). Thousand Oaks: Sage.

Harwit, E. (1995). *China's Automobile Industry: Policies, Problems and Prospects.* New York: M.E. Sharpe.

He, G. (1994). Clarify Goals, Improve Strategies and Invigorate the Machinery and Automobile Industry. *Construction & Design for Project, 3*, 2–5. (in Chinese).

Held, D., McGrew, A., Goldblatt, D., & Perraton, J. (1999). *Global Transformations: Politics, Economics and Culture.* Stanford: Stanford University Press.

Hirst, P., & Thompson, G. (1999). *Globalization in Question: The International Economy and the Possibilities of Governance.* Cambridge, MA: Blackwell Publishers.

Historical Record Office of First Auto Works. (1991). *The Factory History of Fist Auto Work.* Changchun: Jilin Science and Technology Press. (in Chinese).

Historical Record Office of First Auto Works. (1998). *The History of Hongqi Car.* Changchun: First Auto Works. (in Chinese).

Historical Record Office of First Auto Works. (2002). *FAW Yearbook 2001.* Changchun: First Auto Works. (in Chinese).

Hu, C., Chen, X., & Yu, X. (2001). A Comparative Study on the Economic Development of Anhui. *Shanghai Economy Review, 10*, 70–79. (in Chinese).

Huang, Y. (2008, October 20). Zhejiang Province: A Free-Market Success Story. *Business Week*.

Humphrey, J., & Schmitz, H. (2002). How Does Insertion in Global Value Chains Affect Upgrading in Industrial Clusters? *Regional Studies, 36*(9), 1017–1027.

Hurst, W. (2009). *The Chinese Worker after Socialism*. Cambridge, UK: Cambridge University Press.

Hymer, S. H. (1976). *The International Operations of National Firms: A Study of Direct Foreign Investment*. Cambridge, MA: MIT Press.

Iwasaki, I., & Suzuki, T. (2016). Radicalism Versus Gradualism: An Analytical Survey of the Transition Strategy Debate. *Journal of Economic Surveys, 30*(4), 807–834.

Jacobsen, J. K. (1995). Much Ado About Ideas: The Cognitive Factor in Economic Policy. *World Politics, 47*(2), 283–310.

Jiang, Z. (1992). Speed Up the Reform, Opening, and Modernization, Achieve More for Socialism with Chinese Characteristics. *Speech at the 14th National Congress of Communist Party of China*. (in Chinese).

Jiang, Z. (2008, August 24). The Record of the Second Stage of SAIC-VW. *Shanghai Auto News*. (in Chinese).

Jiang, T., & Qiu, K. (1998). The Reform and Opening Shed Lights on Shanghai Automobile Industry. In The Shanghai Automotive Industry (Group) Corporation (Ed.), *Flying High* (pp. 51–58). Shanghai: The Shanghai Automotive Industry (Group) Corporation. (in Chinese).

Jin, J. (1984). Changes of the Economic Position and Role of Shanghai Before and After the Liberation. *Social Sciences, 10*, 14–19. (in Chinese).

Jing, C. (2008). Reflecting the Ownership Institution in China from the Perspective of Central-Local Relationship and State-Owned Assets. *Lanzhou Study Journal, 5*, 98–106. (in Chinese).

Johnson, C. (1982). *MITI and the Japanese Miracle: The Growth of Industrial Policy, 1925–1975*. Stanford: Stanford University Press.

Kaplinsky, R. (2000). Globalisation and Unequalisation: What Can Be Learned from Value Chain Analysis? *Journal of Development Studies, 37*(2), 117–146.

Ke, Q. (1956). Principle and Task of the Shanghai Industry. In Shanghai Planning Committee (Ed.), *Collection of Plans and Reports in Shanghai (1949–1998)* (pp. 575–582). Shanghai Planning Committee: Shanghai. (in Chinese).

Keister, L. A., & Borelli, E. P. (2012). Market Transition: An Assessment of the State of the Field. *Sociological Perspectives, 55*(2), 267–294.

Koeller, C. T. (1995). Innovation, Market Structure and Firm Size: A Simultaneous Equations Model. *Managerial and Decision Economics, 16*(3), 259–269.

Kogut, B. (1984). Normative Observations on the International Value-Added Chain and Strategic Groups. *Journal of International Business Studies, 15*(2), 151–167.

Kornai, J. (1979). Resource-Constrained Versus Demand-Constrained Systems. *Econometrica, 47*(4), 801–819.

Kornai, J. (1986). The Soft Budget Constraint. *Kyklos, 39*(1), 3–30.

Kornai, J. (1992). *The Socialist System*. Princeton: Princeton University Press.

Li, G. (1983). Four Important Steps to Revitalize the Chinese Automotive Industry. In China National Automotive Industry Corporation (Ed.), *China Automotive Industry Yearbook 1983* (pp. 40–50). Beijing: Machinery Industry Press. (in Chinese).

Li, H. (1996). The Basic Ideas of Strategically Restructuring the State-Owned Economy in the Ninth Five-Year Plan. *China Industrial Economy, 7*, 5–8. (in Chinese).

Li, A. (1998). *Family Cars Attracting China*. Beijing: Writers Press. (in Chinese).

Li, G. (2001). Reliance or Self-Reliance, Chinese Automobile Industry in the Opened Environment. *Auto & Society, 4*, 8–11. (in Chinese).

Li, J., & Niu, M. (2003). Chery, a Black Horse in the Chinese Automobile Industry. *Anhui Taxation, 2*, 16–19. (in Chinese).

Liew, L. H. (1995). Gradualism in China's Economic Reform and the Role for a Strong Central State. *Journal of Economic Issues, 29*(3), 883–895.

Lin, J. Y. (2012). *New Structural Economics: A Framework for Rethinking Development*. The World Bank. Washington, DC.

Link, A. N. (1980). Firm Size and Efficient Entrepreneurial Activity: A Reformulation of the Schumpeterian Hypothesis. *Journal of Political Economy, 88*, 771–782.

Liu, Y. (2006). Working and Studying with Minister Lv. In First Auto Works & China Auto News (Eds.), *Lv Fuyuan and Chinese Automotive Industry* (pp. 54–60). Beijing: People's Press. (in Chinese).

Liu, Y., & Zhou, D. (1983). Historical Overview of Anhui's 'Responsibility System' in 1961. *Theoretical Frontline, 179*, 1–11. (in Chinese).

Lu, J. (1998). My Career in Shanghai Automobile Industry. In The Shanghai Automotive Industry (Group) Corporation (Ed.), *Flying High* (pp. 36–44). Shanghai: The Shanghai Automotive Industry (Group) Corporation. (in Chinese).

Lu, F., & Feng, K. (2005). *The Policy Choice to Develop Chinese Automobile Industry with Independent Intellectual Property Rights*. Beijing: Peking University Press. (in Chinese).

Ma, B. (1988). The Historical Ideas and Practices of Shanghai Economic Development. *Shanghai Economic Studies, 5*, 74–79. (in Chinese).

Ma, J., & Liu, H. (2000). *The History and Perspective of the Reform of the State-Owned Enterprises in China*. Beijing: Capital University of Economics and Business Press. (in Chinese).

Mahoney, J. (2000). Path Dependence in Historical Sociology. *Theory and Society, 29*, 507–548.

Mann, J. (1997). *Beijing Jeep: A Case Study of Western Business in China*. Boulder: Westview Press.

McCleery, R. K., & De Paolis, F. (2008). The Washington Consensus: A Post-Mortem. *Journal of Asian Economics, 19*(5–6), 438–446.
McMichael, P. (2003). *Development and Social Change: A Global Perspective*. Thousand Oaks: Pine Forge Press.
Mei, Y., & Feng, K. (2005). Making Cars in Geely: A Report of Self-Reliant Innovation of Geely. *China Soft Science, 11*, 1–10. (in Chinese).
Meng, H. (2003). *Made by Private Enterprises: The Road of Geely and the Situation of Chinese Private Enterprises*. Guangzhou: Guangdong Tourism Press. (in Chinese).
Nan, Y., & Chen, P. (2004, February). Chery Running Through a Narrow Path. *Money China*, 28–33. (in Chinese).
National Planning Committee. (1997). *The Developmental Strategy of Passenger Cars for Chinese Households*. National Planning Committee. (in Chinese).
National Statistics Bureau. (1982). *China Statistics Yearbook 1981*. Beijing: China Statistics Press. (in Chinese).
National Statistics Bureau. (1999). *Major Structure Adjustment and Rapid Economic Growth*. http://www.stats.gov.cn/ztjc/ztfx/xzg50nxlfxbg/200206/t20020605_35960.html. Accessed 10 June 2017. (in Chinese).
Naughton, B. (1994a). *Growing Out of the Plan: Chinese Economic Reform, 1978–1993*. New York: Cambridge University Press.
Naughton, B. (1994b). Chinese Institutional Innovation and Privatization from Below. *The American Economic Review, 84*(2), 266–270.
Nee, V. (1989). A Theory of Market Transition: From Redistribution to Markets in State Socialism. *American Sociological Review, 54*(5), 663–681.
Nee, V. (1992). Organizational Dynamics of Market Transition: Hybrid Forms, Property Rights, and Mixed Economy in China. *Administrative Science Quarterly, 37*(1), 1–27.
Nee, V., & Opper, S. (2012). *Capitalism from Below: Markets and Institutional Change in China*. Cambridge, MA: Harvard University Press.
Nelson, R. R., & Winter, S. G. (1982). *An Evolutionary Theory of Economic Change*. Cambridge, MA: Belknap Press.
Nicholas, T. (2003). Why Schumpeter Was Right: Innovation, Market Power, and Creative Destruction in 1920s America. *The Journal of Economic History, 63*(4), 1024–1058.
North, D. C. (1990). *Institutions, Institutional Change and Economic Performance*. Cambridge, UK: Cambridge University Press.
Oi, J. C. (1992). Fiscal Reform and the Economic Foundations of Local State Corporation in China. *World Politics, 45*(1), 99–126.
Oi, J. C. (1996). *Rural China Takes Off: Incentives for Industrialization*. Berkeley: University of California Press.
Pan, M. (1997). The Statement of the Fortune-Making Secretary. In Zhejiang Historical Archive Committee (Ed.), *Small Commodity and Large Market:*

Memory of the Builders of the Yiwu Small Commodity Market (pp. 101–108). Hangzhou: Zhejiang People's Press. (in Chinese).

Panic, M. (2007). Does Europe Need Neoliberal Reforms? *Cambridge Journal of Economics, 31*(1), 145–169.

Panitch, L. (1995). Globalization and the State. In R. Miliband & L. Panitch (Eds.), *Socialist Register 1994: Between Globalism and Nationalism* (pp. 60–93). London: Merlin Press.

Pereira, A. A. (2008). Whither the Developmental State? Explaining Singapore's Continued Developmentalism. *Third World Quarterly, 29*(6), 1189–1203.

Perraton, J., Goldblatt, D., Held, D., & McGrew, A. (1997). The Globalization of Economic Activities. *New Political Economy, 2*(2), 257–277.

Planning Committee of Shanghai. (1957). *The Report of Shanghai Second Five Year Plan*. Planning Committee of Shanghai. (in Chinese).

Porter, M. E. (1985). *Competitive Advantage*. New York: Free Press.

Qian, Y., & Xu, C. (1993). Why China's Economic Reforms Differ: The M-Form Hierarchy and Entry/Expansion of the Non-State Sector. *Economics of Transition, 1*(2), 135–170.

Rao, B. (1983). United Together, Learn from Experience, and Push the Modernization of the Automobile Industry. In First Auto Works Research Center (Ed.), *The Cradle of the Chinese Automobile Industry* (pp. 22–30). Frist Auto Works: Changchun. (in Chinese).

Rao, B. (1984). Foreword. In China National Automotive Industry Corporation (Ed.), *China Automotive Industry Yearbook 1983* (pp. xi–xiii). Beijing: Machinery Industry Press. (in Chinese).

Rithmire, M. E. (2014). China's "New Regionalism": Subnational Analysis in Chinese Political Economy. *World Politics, 66*(1), 165–194.

Roaf, J., Atoyan, R., Joshi, B., & Krogulski, K. (2014). *25 Years of Transition: Post-Communist Europe and the IMF*. Washington, DC: International Monetary Fund.

Robinson, W. (2001). Social Theory and Globalization: The Rise of a Transnational State. *Theory and Society, 30*, 157–200.

Ronald Burger Strategy Consultants. (2001). *Ten Trends of the Development of China Automobile Industry After WTO*. Ronald Burger Strategy Consultants. (in Chinese).

Ruggie, J. G. (1995). At Home Aboard, Aboard at Home: International Liberalisation and Domestic Stability in the New World Economy. *Millennium: Journal of International Studies, 24*, 507–526.

Sachs, J. (1989). My Plan for Poland. *International Economics, 3*(6), 24–29.

Sachs, J. (1993). *Poland's Jump to the Market Economy*. Cambridge, MA: MIT Press.

Sassen, S. (1996). *Losing Control? Sovereignty in an Age of Globalization*. New York: Columbia University Press.

Scherer, F. (1965). Firm Size, Market Structure, Opportunity and the Output of Patented Inventions. *American Economic Review, 55*(5), 1097–1123.

Schmidt, V. A. (2008). Discursive Institutionalism: The Explanatory Power of Ideas and Discourse. *Annual Review of Political Science, 11*, 303–326.

Schumpeter, J. A. (1942). *Capitalism, Socialism and Democracy.* New York: Harper.

Seo, M., & Creed, W. E. D. (2002). Institutional Contradictions, Praxis, and Institutional Change: A Dialectical Perspective. *Academy of Management Review, 27*(2), 222–247.

Shanghai Zhengda Institute. (2002). *The Strategic Position of Shanghai After WTO.* Beijing: People's Press. (in Chinese).

Sheng, J. (1998). Prelude of the Remodeling Project: Memory of 160 Days in Japan. In Historical Record Office of First Auto Works (Ed.), *Studying in Japan* (pp. 1–14). Changchun: First Auto Works. (in Chinese).

Shi, J., Xie, B., Cheng, H., & Jiang, W. (1987). The "Shanghai Illness" in the 1986 Economy and Its Treatment. *Shanghai Economic Review, 1*, 60–61. (in Chinese).

Shi, J., Wang, W., & Qian, T. (2004). *Private Economy and Institutional Innovation: A Study on Taizhou Phenomenon.* Hangzhou: Zhejiang University Press. (in Chinese).

Sino-Soviet Friendship Association in Jilin and Changchun. (1956). *First Auto Works is Fruit of Friendship between Chinese and Soviet People.* Changchun: Jilin People's Press. (in Chinese).

Social Science Institute of Anhui Province. (1999). Fifty Years, Great Achievement. In Anhui Government (Ed.), *Fifty Years of Anhui* (pp. 3–15). Beijing: China Statistics Press. (in Chinese).

State Council of China. (1994a). *Outlines of the Industry Policy in the 1990s.* State Council of China. (in Chinese).

State Council of China. (1994b). *The Industry Policy of the Automobile Industry.* State Council of China. (in Chinese).

State Council of China. (2004). *The Industry Policy of the Automobile Industry.* State Council of China. (in Chinese).

Steger, M. (2005). Ideology of Globalization. *Journal of Political Ideologies, 10*, 11–30.

Strange, S. (1996). *The Retreat of the State: The Diffusion of Power in World Economy.* New York: Cambridge University Press.

Sturgeon, T., & Florida, R. (2000). *Globalization and Jobs in the Automotive Industry.* Industrial Performance Center, Massachusetts Institute of Technology, Working paper Series. https://ipc.mit.edu/sites/default/files/documents/00-012.pdf. Accessed 10 June 2017.

Sturgeon, T., Van Biesebroeck, G., & Gereffi, G. (2008). Value Chains, Networks and Clusters: Reframing the Global Automotive Industry. *Journal of Economic Geography, 8*, 297–321.

Su, Z. (2007). The Competitiveness of Geely. *China Logistics, 7*, 44–53. (in Chinese).

Su, H., & Ou, Y. (1987). *Anhui Developmental Strategy*. Anhui: Anhui People's Press. (in Chinese).

Symeonidis, G. (1996). *Innovation, Firm Size and Market Structure: Schumpeterian Hypotheses and Some New Themes*. OECD Economics Department Working Papers, No. 161.

Taizhou Statistics Bureau. (2005). *Taizhou Statistics Yearbook 2005*. Beijing: China Statistics Press. (in Chinese).

Tang, L. (2001). The Changes of Strategic Ideas of the Central Committee of CPC on Shanghai after the Establishment of New China. *Party and Government Forum, 6*, 18–20. (in Chinese).

Tao, J. (2005). Chery, Asking Profit from Human Resources. *Chinese Times, 1*, 32–35. (in Chinese).

Technology Center of Shanghai Automotive Industry. (2002). *The Discussions on Improving the Capacity of Self-Reliant Innovation in SAIC Group*. Shanghai: Technology Center of Shanghai Automotive Industry. (in Chinese).

Thelen, K. (1999). The Historical Institutionalism in Comparative Politics. *Annual Review of Political Science, 2*, 369–404.

Thun, E. (2006). *Changing Lanes in China: Foreign Direct Investment, Local Governments, and Auto Sector Development*. New York: Cambridge University Press.

Vernon, R. (1971). *Sovereignty at Bay*. New York: Basic Books.

Wade, R. (1990). *Governing the Market: Economic Theory and the Role of the Government in East Asian Industrialization*. Princeton: Princeton University Press.

Wade, R. (1996). Globalization and Its Limits: Reports of the Death of the National Economy Are Greatly Exaggerated. In B. Suzanne & R. Dore (Eds.), *National Diversity and Global Capitalism* (pp. 60–88). New York: Cornell University Press.

Walder, A. (1995). Local Government as Industrial Firms. *American Journal of Sociology, 101*(2), 263–301.

Wallerstein, I. (1974). *The Modern World-System, Vol. I: Capitalist Agriculture and the Origin of the European World-Economy in the Sixteenth Century*. New York: Academic Press.

Wallerstein, I. (1980). *The Modern World-System, Vol. II: Mercantilism and the Consolidation of the European World-Economy, 1600–1750*. New York: Academic Press.

Wallerstein, I. (1989). *The Modern World-System, Vol. III: The Second Era of Great Expansion of the Capitalist World-Economy, 1730–1840s*. New York: Academic Press.

Walter, A. (2006). From Development to Regulatory State? Japan's New Financial Regulatory System. *The Pacific Review, 19*(4), 405–428.

Wang, L. (1989). *The Discourse of Responsibility System in Anhui: 1961 to 1978*. Beijing: Kunlun Press. (in Chinese).

Wang, Y. (1996a). The Car Industry: Driving Anhui to the New Century (First Part). *Decision Reference, 2*, 6–8. (in Chinese).
Wang, Y. (1996b). The Car Industry: Driving Anhui to the New Century (Second Part). *Decision Reference, 3*, 6–8. (in Chinese).
Wang, X. (1998). From 'Specimen' in the Planned Economy to 'Red Flag' in the Market Economy. *Outlook, 17*, 30–32. (in Chinese).
Wang, X. (2003). Chery, the State Engine and the Market Path. *New Economy, 9*, 38–41. (in Chinese).
Wang, Z. (2003). The Evolution and Prospect of the Economic Development Strategy of Shanghai. In A. Wang (Ed.), *The Development of Shanghai and the Capacity of Governance* (pp. 57–77). Shanghai: Wenhui Press. (in Chinese).
Weber, M. (1978). *Economy and Society: An Outline of Interpretive Sociology*. Berkley: University of California Press.
Wei, X. (2008). *Taizhou People Created the Stock Cooperation System*. http://www.tzdsw.net.cn/newsinfo.asp?ID=477. Accessed 10 June 2017. (in Chinese).
Weik, E. (2011). Institutional Entrepreneurship and Agency. *Journal for the Theory of Social Behaviour, 41*(4), 465–481.
Weiss, L. (1998). *Myth of the Powerless State*. Ithaca: Cornell University Press.
Whiting, S. (2001). *Power and Wealth in Rural China: The Political Economy of Institutional Change*. New York: Cambridge University Press.
Williamson, J. (1990). *Latin American Adjustment: How Much Has Happened?* Washington, DC: Institute for International Economics.
Williamson, O. E. (1993). Transaction Cost Economics and Organization Theory. *Industrial and Corporate Change, 2*(2), 107–156.
Winter, S. G. (1984). Schumpeterian Competition in Alternative Technological Regimes. *Journal of Economic Behavior and Organization, 5*, 287–320.
Woo, W. T. (1999). The Real Reasons for China's Growth. *The China Journal, 41*, 115–137.
Wu, Y. (2008, October). *Fiscal Equalization will Help Prevent Corruptions*. Oriental Outlook. http://news.sohu.com/20081028/n260297950.shtml. Accessed 10 June 2019. (in Chinese).
Wu, Q. (2009). *Unusual Early History of the Chinese Automobile Industry Memorized by the Father of Santana*. http://auto.gasgoo.com/News/2009/01/07101532153249138454.shtml. Accessed 10 June 2017. (in Chinese).
Xiang, B. (2008). *The Right Developmental Way Should Fit Local Reality*. http://www.tzdsw.net.cn/newsinfo.asp?ID=477. Accessed 10 June 2017. (in Chinese).
Xu, Y. (1983). Reflection and Prospect. In First Auto Works Research Center (Ed.), *The Cradle of the Chinese Automobile Industry* (pp. 1–13). Changchun: First Auto Works. (in Chinese).
Xu, C. (1986). Thoughts on the Developmental Strategy of the Economy of Anhui. *Theory Construction, 2*, 23–30. (In Chinese).

Xu, J. (2004). Businessman with Red Hat of Anhui in Trouble. *South Reviews, 4*, 53–55. (in Chinese).

Yan, G. (2003). *The Santana Model Witnessed the Growth of the Chinese Automobile Industry*. http://auto.sina.com.cn/news/2003-07-28/42011.shtml. Accessed 10 June 2017. (in Chinese).

Yang, W. (2005). The Historical Contribution of the Northeast Old Industrial Base to the New China. *References of Economic Study, 91*, 24–28. (in Chinese).

Yang, M. (2008). The Legend of Chery. *Decision-Making, 9*, 21–23. (in Chinese).

Yao, X. (1994). *Fifteen Years of Shanghai in Reform*. Shanghai: Shanghai Social Science Academy Press. (in Chinese).

Yeung, H. W. (2000). State Intervention and Neoliberalism in the Globalizing World Economy: Lessons from Singapore's Regionalization Programme. *The Pacific Review, 13*(1), 133–162.

Yeung, H. W., & Coe, N. M. (2015). Toward a Dynamic Theory of Global Production Networks. *Economic Geography, 91*(1), 29–58.

Yi, B., & Lin, S. (2001). The Industrialization History of the Northeast. *Northeast Asia Forum, 4*, 54–64. (in Chinese).

Yin, T. (1994). Policy of Domestic Content Requirement and Import Regulations. In Ministry of Machinery Industry of China (Ed.), *Material Collection of PSE'94* (pp. 41–45). Beijing: Ministry of Machinery Industry of China. (in Chinese).

Yuan, W., & Xie, D. (2002). The Fragile Geely. *South Reviews, 3*, 53–55. (in Chinese).

Zhan, Y. (2008). An Analysis on the Dialectic Relationship Between Strategic Ally and Enterprise Innovation. *Studies in Dialectics of Nature, 24*(3), 68–72. (in Chinese).

Zhang, W. (2004). *Build Geely as the Chinese Version of Hyundai in Korea*. http:// auto.sohu.com/ 20041120/n223092912.shtml. Accessed 10 June 2017. (in Chinese).

Zhang, Z. (2005). *Top Background Information: Significant Person in the Chinese Economy*. Shantou: Shantou University Press. (in Chinese).

Zhang, S., & Chen, J. (2005). *Chery Auto Breaks the Myth: A Forced Path of Self-reliant R&D*. http://ah.anhuinews.com/system/2005/11/16/001391506.shtml. Accessed 10 June 2017. (in Chinese).

Zhang, R., & Yang, Y. (2006). Zhejiang Merchants: Growth Background, Group Characters, and Developmental Trend. *Business Economics and Administration, 6*, 3–7. (in Chinese).

Zhang, B., Yao, F., Zhang, J., & Jiang, L. (2004). *The Technological Transfer from Soviet Union to China, 1949 to 1966*. Jinan: Shandong Education Press. (in Chinese).

Zhang, M., Xie, S., & Guo, Y. (2005). *Introduction of New Taizhou Merchants*. Changchun: Jilin University Press. (in Chinese).

Zhao, X. (1983). Concentrate to Help the Remodeling in First Auto Works. In First Auto Works Research Center (Ed.), *The Cradle of the Chinese Automobile Industry* (pp. 14–16). Changchun: First Auto Works. (in Chinese).

Zhao, R. (1995). Five Advantages Sectors in the Machinery Industry of Anhui. *China Electronics and Machinery Industry, 10*, 15. (in Chinese).

Zhao, X. (2006). Observing Chery from Outside: Analyze the Development of Chery from a Perspective of Patent. *China Invention & Patent, 4*, 52–55. (in Chinese).

Zhao, X., & Zhang, L. (1999). Decentralization Reforms and Regionalism in China: A Review. *International Regional Science Review, 22*(3), 251–281.

Zhejiang Statistics Bureau. (2006). *Zhejiang Statistics Yearbook 2006*. Beijing: China Statistics Press. (in Chinese).

Zhejiang Statistics Bureau. (2007). *Private Economy Become Critical Engine of Socio-economic Development of Zhejiang*. http://www.stats.gov.cn/ztjc/ztfx/dfxx/200712/t20071225_33874.html. Accessed 10 June 2017. (in Chinese).

Zheng, Y. (1994, August 9). Critique on the Car Civilization. *Guangming Daily*. (in Chinese).

Zheng, Y. (1996). *Debate on Cars*. Beijing: Economic Science Press. (in Chinese).

Zheng, Y. (1998). My Memory on the Project of 30 Thousand Cars in First Auto Works. In Historical Record Office of First Auto Work (Ed.), *The History of Hongqi Car* (pp. 212–227). First Auto Works: Changchun. (in Chinese).

Zheng, M. (2000). The Expansion of Japanese Capital in the Occupied Northeast. *The Frontline of Social Science, 6*, 183–190. (in Chinese).

Zheng, Z. (2007). *Li Shufu, Automobile "Maniac": Decoding the First Private Carmaker*. Beijing: China Citic Press. (in Chinese).

Zhou, Z. (1998). Explore the Developmental Advantages of Anhui Economy. *Economic Theory and Economic Management, 3*, 61–65. (in Chinese).

Zhu, J. (1988). Short of Resource but Rich in Economy: Some Thoughts on Strategy of Economic Development in Zhejiang. *Exploration, 2*, 16–33. (in Chinese).

Zhu, Y. (2004). *An Investigation of Businessman with Red Hat in Wuhu*. http://news.xinhuanet.com/report/2004-02/22/content_3181726.htm. Accessed 10 June 2017. (in Chinese).

Zhu, Q. (2006). The Vigor of Geely. *Ceocio China, 9*, 51–59. (in Chinese).

Zhu, J., Gan, H., et al. (1999). *Shanghai General History (Volume 12): Modern Economy*. Shanghai: Shanghai People's Press. (in Chinese).

Zun, L. (2007). *Innovation of Chery*. Beijing: China Citic Press. (in Chinese).

Index[1]

[1] Note: Page numbers followed by "n" refer to notes.

Q. Feng, *Variety of Development*,
https://doi.org/10.1007/978-981-10-5912-4

www.ingramcontent.com/pod-product-compliance
Lightning Source LLC
LaVergne TN
LVHW082319171026
850499LV00004B/90
* 9 7 8 9 8 1 1 0 5 9 1 1 7 *